Improving CAD Designs wit

Autodesk Fusion 360

A project-based guide to modelling effective
parametric designs

Kevin Michael Land

BIRMINGHAM—MUMBAI

Improving CAD Designs with Autodesk Fusion 360

Group Product Manager: Rohit Rajkumar

Publishing Product Manager: Vaideeshwari Muralikrishnan

Senior Content Development Editor: Rakhi Patel

Technical Editor: K Bimala Singha

Copy Editor: Safis Editing

Project Coordinator: Aishwarya Mohan

Proofreader: Safis Editing

Indexer: Subalakshmi Govindhan

Production Designer: Shankar Kalbhor

DevRel Marketing Coordinators: Anamika Singh, Namita Velgekar, and Nivedita Pandey,

First published: September 2023

Production reference: 1030823

Published by Packt Publishing Ltd.
Grosvenor House
11 St Paul's Square
Birmingham
B3 1RB, UK.

ISBN 978-1-80056-449-7

www.packtpub.com

I want to thank my wife, Katia, and my daughter, Minako, for their patience, love, and support while I worked multiple nights writing this book. Thank you to my mom, June; my sister, Cathy; and my brother, Bob for always being there for me. Love to you all.

To the memory of my father, William Kevin Land, thank you for all the sacrifices that you made and for working so hard at a job that you hated, allowing me to do something that I love. I miss you every day.

– Kevin Land

Contributors

About the author

Kevin Michael Land is a certified AutoCAD instructor and has worked in a variety of disciplines throughout his 25-year career within electrical, landscape, piping, mechanical, and architectural firms. He has used a variety of software throughout the years, ranging from AutoCAD, MicroStation, Revit, SolidWorks and, his personal favorite, Fusion 360. He is a freelance instructor who teaches at education institutes such as **Future Media Concepts** (**FMC**) and NetCom Learning, and tutors students via Wyzant.com. He currently works at Walt Disney World in Lake Buena Vista, FL as a designer in the Ride and Show Design Services department and loves working on some of his favorite rides. He holds a Bachelor of Science in applied technology from Wilmington University and an AutoCAD R13 certificate from the **Computer Processing Institute** (**CPI**). In his free time, he likes to spend time playing with his 11-year-old daughter, 3D printing objects via his Flux Delta 3D printer, and making low-budget movies.

To the Packt Publishing team, thank you all for your guidance and help in publishing this book. You are all rockstars and are amazing at your jobs. To anyone who has ever thought about writing your own book, reach out to the Packt team as they have the knowledge and know how to make it happen.

About the reviewer

Zain Razzaq is a passionate mechanical design engineer with over five years of experience in the research and development industry. He is an Autodesk Certified Professional in Design for Manufacturing using Fusion 360, and his expertise lies in designing for manufacturing, ergonomic design, and generative design.

Zain is currently a design and applications engineer for Tecnica Inc., where he is responsible for developing industrial 3D printers and optimizing existing designs. He is also a mechanical design/ thermal simulations engineer at Whizz Systems, where he uses his skills to improve the performance and reliability of products.

In addition to his work in industry, Zain is also an active educator. He has taught 3D design undergraduate courses to senior-year students at the University of Management and Technology. He is also a frequent speaker at industry events, where he shares his knowledge and expertise with others.

Zain is a highly skilled and experienced mechanical design engineer with a deep understanding of the latest design technologies. He is passionate about his work and is always looking for new ways to improve the design and manufacturing of products. He is also a dedicated educator who is committed to sharing his knowledge with others.

Table of Contents

Part 1: Simple, Fun Projects for Around the Home

1

2

3

4

5

Part 2: Bicycle Water Bottle Holder Project

6

7

8

12

Modeling a Scary Tealight Ghost   383

13

Using Form and Solid Modeling to Create a Cushioned Chair   421

Part 4: Working with 2D and 3D Scanned Images

14

Using a Scanned Image to Create a 3D Model   481

15

Preface

Welcome, fellow makers and soon-to-be makers, to *Improving CAD Designs with Autodesk Fusion 360*! Fusion 360 has grown to be a favorite tool used by CAD, CAM, CAE, and PCB creative professionals and average household makers. In this book, we will explore ways to design different products for around-the-home use and improve upon these designs while we create. If you have access to a 3D printer, all these parts can be 3D printed for use and fun at home. I created this book as a way for beginner makers to get an idea of what makes a part, how to design your own parts, and how to improve upon your designs.

Who this book is for

This book is intended for all beginner designers who would like to learn through experimentation rather than following a direct straight-to-the-point tutorial. This is not a book for mechanical engineers but more for designers, woodworkers, 3D printing enthusiasts, and hobbyists who love to create.

What this book covers

Chapter 1, *Working within the Design Workspace*, introduces the book with a brief overview of the most commonly used Fusion 360 user interface. This chapter will show the differences between the variety of other workspaces within Fusion 360 and the different tools located within each.

Chapter 2, *Planes, Sketches, Constraints, and Parametric Dimensions*, details the most used tools that make Fusion 360 a powerful design program and what these tools do.

Chapter 3, *Project Building Basics*, gives you an overview of where Fusion 360 stores projects in the cloud and why they are not located on your machine. This chapter shows you how to load files to the cloud and how to share projects with other users. The chapter also goes over how to work with multiple file versions and how to open previous versions.

Chapter 4, *Creating a Customizable S-Hook*, offers a look into two different ways to create an S-Hook hanger in Fusion 360 using the **Sweep** and **Pipe** tools. This chapter also demonstrates what Rule #1 is and how to create components.

Chapter 5, *Designing Decorative Doorknobs*, explains three different ways to create a simple doorknob using either addition, subtraction, or profile revolution. This chapter demonstrates various ways to create the same model.

Chapter 6, Designing a Simple Bottle Holder, explores how to plan and design a bottle holder that can be attached to a bicycle handlebar. This chapter demonstrates how to use calipers and how to take measurements of bicycle handlebars.

Chapter 7, Creating a Bike Reference Model, teaches you how to take measurements from a paper-and-pencil sketch and bring those measurements into Fusion 360. This chapter demonstrates how to create a reference model of the handlebars to be used with the bottle holder model.

Chapter 8, Creating a Bottle Reference Model, is centered around the example of averaging out existing measurements from multiple water bottles to best determine the size of the reference bottle to use to create the parametric bottle holder model.

Chapter 9, Building the Bottle Holder, goes through the steps for creating the water bottle holder by using the bicycle handlebars and water bottle as reference models. This chapter demonstrates how the bottle holder uses the water bottle as a reference model to gather the necessary dimensions.

Chapter 10, Improving the Bottle Holder Design, explores how to improve the bottle holder design by correcting errors to create a better-designed model. This chapter explores how to use the timeline to go back and make corrections to an existing model.

Chapter 11, The FORM Environment, explains how to work with mesh objects in the Form environment. This chapter demonstrates how models are more artistic and act similarly to working with clay.

Chapter 12, Modeling a Scary Tealight Ghost, is all about the surface environment. The tutorial in this chapter demonstrates how to use surface tools such as **Patch** and **Thicken** to create a model of a tealight ghost.

Chapter 13, Using Form and Solid Modeling to Create a Cushioned Chair, contains a tutorial on how to combine working in both the form and solid environments to create a gaming chair. This chapter demonstrates how to sculpt the chair in the Form environment and then switches over to a solid modeling environment to build the interior structure.

Chapter 14, Using a Scanned Image to Create a 3D Model, outlines a tutorial on how to take a scanned image into Fusion 360, scale it to the correct size, and then build a model from it.

Chapter 15, Modeling a Bottle Topper, demonstrates how to import the mesh model of a person's upper body, fix any holes and errors within the model, and then turn the model into a solid so that it can be edited further.

To get the most out of this book

To get the most out of this book, you will need to have a basic understanding of PC or Mac commands. No prior knowledge of CAD is necessary for the beginner projects but, if you'd like to skip ahead, it is recommended that you have some basic CAD knowledge for the projects in later chapters. It is highly recommended that you use a three-button mouse for all projects to take advantage of floating pop-up menus.

Software/hardware covered in the book	Operating system requirements
Fusion 360	Windows or macOS

There is a paid version of Fusion 360, which gives you the ability to save more files without having to archive older files, but it is not necessary to use for this book. When the 10-file limit is reached, simply archive the older files, which will not delete them but will allow for more saved files to be created.

I would recommend when you finish a chapter to go back and re-try the tutorial without using the instructions and see whether you can recreate the file or create something new. The best advice I can give is to always play and experiment with the files. Experiment with different ways of working and see what happens. Practice, practice, practice!

Download the example code files

You can download the sample designs for this book from GitHub at `https://github.com/PacktPublishing/Improving-CAD-Designs-with-Autodesk-Fusion-360`. If there's an update to the code, it will be updated in the GitHub repository.

We also have other code bundles from our rich catalog of books and videos available at `https://github.com/PacktPublishing/`. Check them out!

Conventions used

There are a number of text conventions used throughout this book.

`Code in text`: Indicates code words in text, database table names, folder names, filenames, file extensions, pathnames, dummy URLs, user input, and Twitter handles. Here is an example: "Let's stick with Rule #1 and create a new component and name it `Doorknob_Revolve_Addition`."

Bold: Indicates a new term, an important word, or words that you see onscreen. For instance, words in menus or dialog boxes appear in **bold**. Here is an example: "Click on the **CREATE** panel dropdown and then click on **Extrude**."

> Tips or important notes
> Appear like this.

Get in touch

Feedback from our readers is always welcome.

General feedback: If you have questions about any aspect of this book, email us at customercare@ packtpub.com and mention the book title in the subject of your message.

Errata: Although we have taken every care to ensure the accuracy of our content, mistakes do happen. If you have found a mistake in this book, we would be grateful if you would report this to us. Please visit www.packtpub.com/support/errata and fill in the form.

Piracy: If you come across any illegal copies of our works in any form on the internet, we would be grateful if you would provide us with the location address or website name. Please contact us at copyright@packt.com with a link to the material.

If you are interested in becoming an author: If there is a topic that you have expertise in and you are interested in either writing or contributing to a book, please visit authors.packtpub.com.

Share Your Thoughts

Once you've read *Improving CAD Designs with Autodesk Fusion 360*, we'd love to hear your thoughts! Scan the QR code below to go straight to the Amazon review page for this book and share your feedback.

https://packt.link/r/180056449X

Your review is important to us and the tech community and will help us make sure we're delivering excellent quality content.

Download a free PDF copy of this book

Thanks for purchasing this book!

Do you like to read on the go but are unable to carry your print books everywhere?

Is your eBook purchase not compatible with the device of your choice?

Don't worry, now with every Packt book you get a DRM-free PDF version of that book at no cost.

Read anywhere, any place, on any device. Search, copy, and paste code from your favorite technical books directly into your application.

The perks don't stop there, you can get exclusive access to discounts, newsletters, and great free content in your inbox daily

Follow these simple steps to get the benefits:

1. Scan the QR code or visit the link below

https://packt.link/free-ebook/978-1-80056-449-7

2. Submit your proof of purchase
3. That's it! We'll send your free PDF and other benefits to your email directly

Part 1:
Simple, Fun Projects for
Around the Home

In this part, you will be introduced to the Fusion 360 user interface and understand how the tools work, how the interface changes while you use different tools, and how to set up a simple project. You will then create a basic part that can be 3D-printed if you have access to a 3D printer.

This part has the following chapters:

- *Chapter 1, Working within the Design Workspace*
- *Chapter 2, Planes, Sketches, Constraints, and Parametric Dimensions*
- *Chapter 3, Project Building Basics*
- *Chapter 4, Creating a Customizable S-Hook*
- *Chapter 5, Designing Decorative Doorknobs*

1
Working within the Design Workspace

Fusion 360 is a fantastic 3D parametric, mesh design, and **CNC** program that you can use to create almost anything imaginable. In this chapter, first, you will get an idea of how to navigate the workspace and learn where most of the tools are located. Then, you will learn the difference between top-down and bottom-up design before understanding the various pricing methods of Fusion 360 and how to get the hobbyist license. The main goals of this chapter are to give you a basic understanding of Fusion 360 and help you locate tools, as well as show you where to find more help when you run into any problems.

In this chapter, we're going to cover the following main topics:

- The basics of the **user interface (UI)**
- Exploring the various design approaches
- Various pricing methods of Fusion 360

Technical requirements

There are no technical requirements for this chapter, but it is highly recommended that you use a three-button mouse while using Fusion 360.

Please take a look at the technical requirements for installing Fusion 360 on the Autodesk website: knowledge.autodesk.com/support/fusion-360.

Basics of the UI

The Fusion 360 UI has been laid out to resemble most current-day program interfaces, with a toolbar at the top holding the most used commands, a data panel to the left, and a large workspace in the center. The UI will change as you select certain buttons and change to other workspaces. Like most programs, if you hover your mouse over a button and wait, you can see what the command does, and by hitting *Ctrl* + /, you can pull up the help menu for that command. If that doesn't work, try restarting Fusion 360 and trying again.

The following figure (*Figure 1.1*) shows a brief overview of the Fusion 360 UI, with the data panel open on the left and a gestures palette (see **6**) open on the right. The gestures command can be used with *Shift* + right-click, then mousing over the location of the command you wish to use:

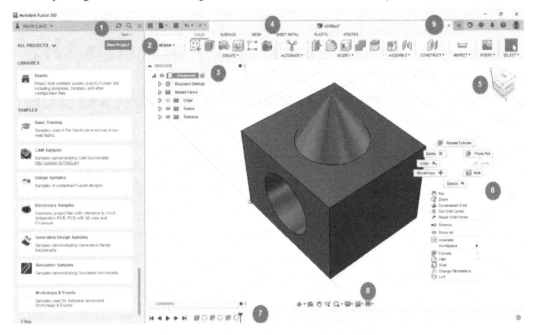

Figure 1.1 – The user interface of Fusion 360

This chart legend will help you understand the location of a button and what it is used for. We will look at these commands in more depth as we use our projects:

Bubble #	Icon	Name	Description
1		Data panel	Stores your designs and manages projects. Click once to open and again to close.
2	DESIGN ▾	Workspace	Each workspace changes the ribbon at the top of the screen. They are separated into DESIGN, GENERATIVE DESIGN, RENDER, ANIMATION, SIMULATION, MANUFACTURE, and DRAWING. Some workspaces may have limited capabilities, depending on your subscription.

Bubble #	Icon	Name	Description
3	◀◀ BROWSER	Browser	Shows a list of all assemblies, bodies, sketches, and other tools that you used while creating your part.
4	SOLID SURFACE MESH / CREATE ▼	Toolbar	The ribbon will change as the workspace changes. It is separated into multiple tabs to make locating commands easier.
5		View cube	Orbit by left-clicking and dragging over the cube. Left-click once on the corners or faces to view your 3D model.
6	*Shift* + right-click	Marking menu	Typically known as the gestures menu, you can invoke certain commands by quickly holding down *Shift* + right-clicking and dragging your mouse to a specific location to start the command.
7	⏮ ◀ ▶ ▶▶ ⏭	Timeline	Shows the list of command operations in order. This order can be manipulated by clicking and dragging on icons within certain rules. You can right-click on the icons to make changes to that command at that moment in time.
8		Navigation bar	Contains the locations for **Zoom**, **Pan**, and **Orbit**. Changes the look of the model and the appearance of the UI.
9		Application bar	The application bar contains items to the left of the screen, such as the data panel, **File**, **Save**, **Undo**, and **Redo**, as well as items to the right, such as **Extensions**, **Job status**, **Notifications**, **Help**, and **My Profile**.

Now that we understand where items are placed within the Fusion 360 UI, we will now take a look at one of the most important and frequently used buttons that I look for when first learning a new software, which is the help button. The help button is essential for learning any new software, and Fusion 360 has an enhanced feature to make it even easier to learn while using help.

The help menu

One of the main reasons I chose to create a project-based learning book is due to the amazing existing help selections that have been created for Fusion 360. There are so many options out on the internet, either on YouTube or directly on the Autodesk website, from which you can locate any command and learn its function.

When learning any new software, it is important to locate the help menu as it can be difficult to remember what a command does or even where it is located. We will only discuss the most important and frequently used ones but keep this area in mind whenever you have more questions about Fusion 360.

You can quickly get to the help menu by going to the top right of the UI and left-clicking on the *question mark* icon. You can also access this menu by mousing over a button in the ribbon and hitting *Ctrl +* / to quickly open the **Learning Panel** area:

Figure 1.2 – The location of the help menu

The help menu contains a few different menu options within it, such as the redesigned **Learning Panel** area (*Figure 1.3*), which, if left open, will update as you click on different button commands in the UI:

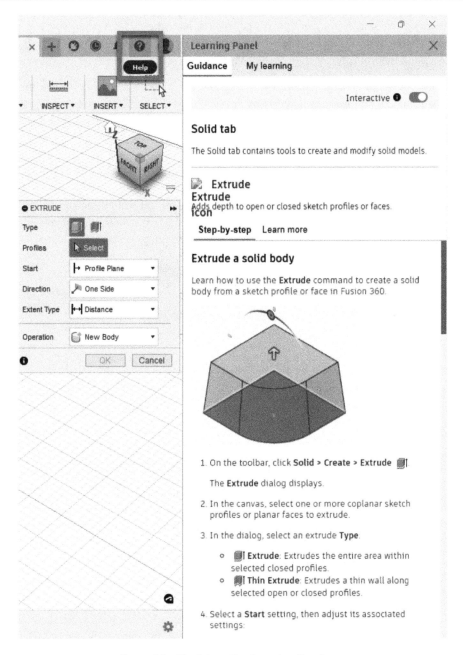

Figure 1.3 – The interactive Learning Panel area

The help menu of Fusion 360 is a great tool to have when learning this program. There is a lot of great information located within here and I'd recommend keeping it out when you're first learning the commands and their purpose.

Another great resource for gathering ideas and seeing how other Fusion 360 creators have built their projects is the Autodesk Community Gallery: `https://www.autodesk.com/community/gallery`. Within this web page, you can view pictures and sometimes the 3D models of the projects themselves and see how other users built their products.

Preferences

If you are used to working in another program or would like to change something you don't like, the first stop is usually in **Preferences**. This is where you can set up Fusion 360 the way you want it to work. You can find **Preferences** to the right of your screen by clicking on the *person* icon:

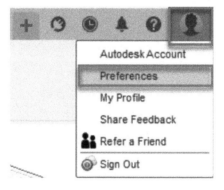

Figure 1.4 – Preferences

Within the **Preferences** panel, under the **General** tab, you can set which default Z orientation (shown in red in *Figure 1.5*) you would like to work with. This is important as the default Z direction can change between computer programs such as Blender or SolidWorks. I typically like mine facing up rather than toward my screen. If you are completely new to the X, Y, and Z axes, don't worry – I will explain them while we are working on the projects.

> **Axis orientation note**
>
> There are three main axes in Fusion 360, which are the X-axis, the Y-axis, and the Z-axis. The purpose of these axes is to give you a sense of direction when modeling in 3D space. Refer to *Chapter 2* for more information on how they can be used in Fusion 360.

The **Pan, Zoom, Orbit shortcuts** area (shown in green in *Figure 1.5*) is typically the next thing to customize. You can set this to either the default application of Fusion 360 or Alias, Inventor, SolidWorks, Tinkercad, or PowerMILL. I have mine set to Tinkercad as it's close to AutoCAD, which I've worked with for over 20 years; *old habits are hard to break*.

The last setting here is **Reverse zoom direction** (shown in yellow in *Figure 1.5*). If you're using a three-button mouse, by default, scrolling in with your mouse will zoom you in toward whatever object you have on screen. If you check this box, it will reverse that direction:

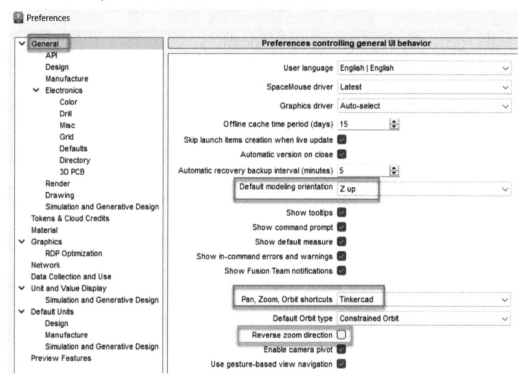

Figure 1.5 – The Preferences panel

There are many other preferences to search through, too many for this book, but I recommend looking through all of these to become familiar with settings that you may want to tweak at some point. The other recommendation I would say is to check out the **Preview Features** section (see *Figure 1.6*). This area shows some new tools that are still in beta version and lets you experiment with them before they become part of Fusion 360 (or don't):

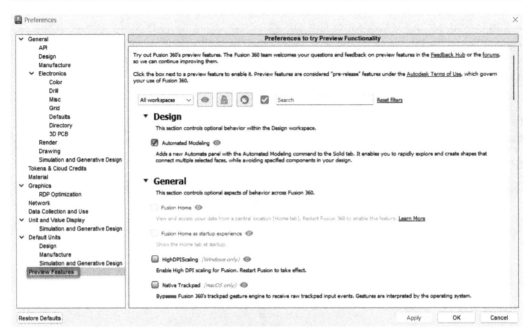

Figure 1.6 – The Preferences panel

The data panel

The data panel is where all of your projects will be stored, as well as any subsequent drawings, PDFs, images, or any other project information that may be relevant:

Figure 1.7 – The data panel

It is best to keep things organized, as you would in a typical project, by hitting the **New Project** button (shown in red in *Figure 1.8*) to create a project folder. This folder can be shared with other designers working on your project so that users can collaborate without having to create multiple drawings.

If you have trouble locating a project file, you can search for a particular name by clicking the magnifying glass icon toward the top right (shown in green in *Figure 1.8*):

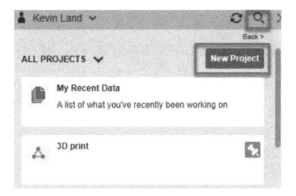

Figure 1.8 – The New Project button's location

If you scroll to the bottom of the data panel, you will find links to some sample projects that you can practice with and see how they were all put together:

Figure 1.9 – The location of sample projects, which can be found toward the bottom of the data panel

Workspaces

Fusion 360 has seven different workspaces. Each will change how the ribbon will look and allow you to switch to a different design space to suit another design purpose. Most of the time, you will do your creating in the **DESIGN** workspace, which is also the default space that Fusion 360 starts with. Once you have finished your design, you may want to move on to testing it within the **SIMULATION** workspace or create a quick render or animation to send off to a client through the **RENDER** or **ANIMATION** workspace, respectively, or send your design for manufacturing on a CNC machine through the **MANUFACTURE** workspace. Each workspace will take some time to work in and some are harder than others to learn as some may require engineering knowledge of weights, pressure, and materials. Do not be afraid of these workspaces, though; experiment and learn by taking a look at each workspace and seeing what can be done within each:

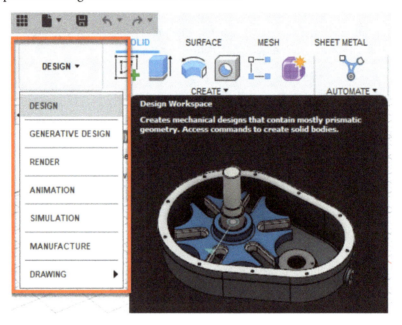

Figure 1.10 – The seven different workspaces

Toolbar

The toolbar, like most Windows-based programs, containerizes button locations into tabs such as **SOLID, SURFACE, MESH, SHEET METAL, PLASTIC,** and **UTILITIES**, and then into separate groups such as **CREATE, AUTOMATE** (this only appears in paid versions), **MODIFY, ASSEMBLY, CONSTRUCT, INSPECT, INSERT,** and **SELECT**. We will go over each as we work through a variety of projects in this book. For now, get a basic understanding of where buttons are located and how each tab will change the group of buttons within it:

Figure 1.11 – The DESIGN toolbar

The toolbar buttons can be reorganized within their grouping by clicking and dragging a button to a new location. If you want to add a new button to the toolbar, do the following:

1. Hover over a button that you would like to add.

2. Go to the right of the button and click on the three vertical dots.

3. Check the **Pin to Toolbar** box.

4. Now, you can rearrange its location by clicking and dragging the button icon on the toolbar:

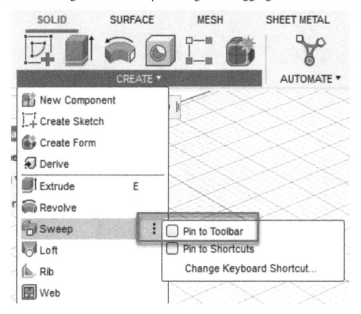

Figure 1.12 – The location of the Pin to Toolbar checkbox

While there are various shortcut keys that you can use in Fusion, all of which we will explore while working on our projects, most of your time will be spent locating tools within the toolbar. It's a good idea to get familiar with locating the most used tools and customizing the toolbar to help you work faster.

Now that we are familiar with the basics of the UI, such as the help menu, the **Preferences** panel, and the toolbar, we will now go over two different ways to design products. One is the top-down approach, wherein a completely new design is created, while the other is the bottom-up approach, wherein we use existing products to update them to fit new needs.

Exploring the various design approaches

There are various ways to create a simple object in Fusion 360. I will go over a few different ways to draw, each of which has its benefits and downsides. It's up to you, as a designer, to choose which approach works best for you and the next changes you will make to the model. The main questions to think about when designing are how quickly the object can be created, how easy it would be to modify it if there were a design change, and how accurate the model is.

The top-down design method

Designing from the top down means that you can create custom parts within Fusion 360 and design them to fit other custom parts. You take the time to plan out your model before anything is created. This is a more structured approach. These custom parts can control other parts, as shown in *Figure 1.13*. The custom legs control the design of the custom tabletop:

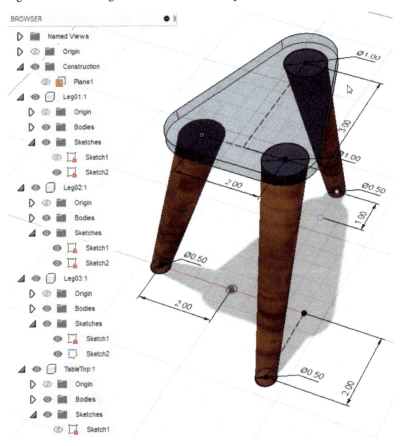

Figure 1.13 – A top-down approach to modeling that involves taking a custom table and using its legs to control the design of the tabletop

As the legs change to different locations, the table will change as well (see *Figure 1.14*). The parts that you design can be exported to other models if need be:

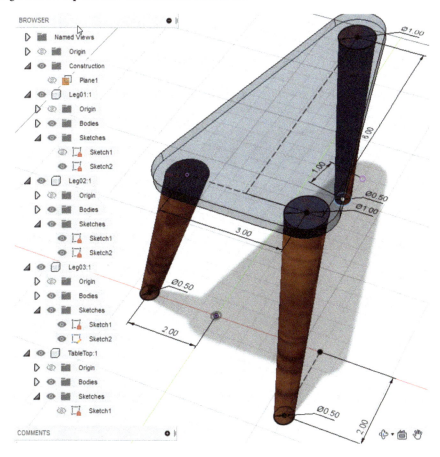

Figure 1.14 – A design change controlled by the table legs' dimensions

This is a very powerful design method but requires some forethought as to what part you want to drive other parts. The best way to learn which design method is best is through experimenting.

The bottom-up design method

Design from the bottom up involves starting with smaller pieces and building to a larger project. This takes on more of a creative side to design since you start from an existing part and create from there. An example of this is when you know of the design goal and have all the main parts to insert into Fusion, such parts as from another designer on your team or parts from McMaster-Carr. You pull all these parts into Fusion 360 and then start assembling them to build the main assembly. Then, you design some other parts to fit this assembly and manipulate them as you design. You will end up with more separate models in your library, but these models can be used across other designs:

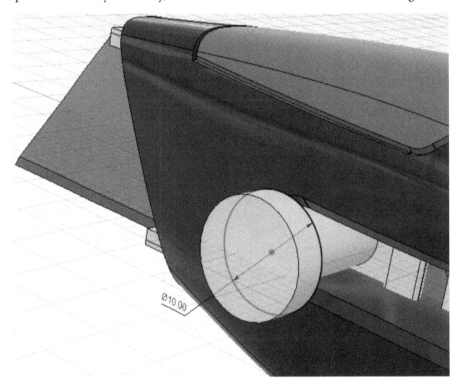

Figure 1.15 – A bottom-up approach to modeling and adding to an existing model

Some prefer one way to design over the other but a blend of both can often provide the best results. You will often find that starting from the top down is the easier way to start, but as you grow as a modeler, using both methods will make you a much stronger designer. Being able to switch how you work and not be confined to a certain methodology will make your models much more efficient to work with:

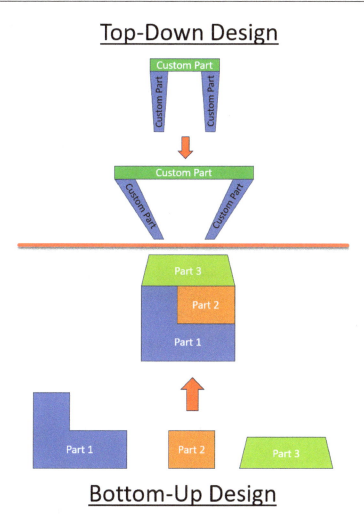

Figure 1.16 – The top-down and bottom-up design workflow

Now that we've learned about the UI and the difference between top-down and bottom-up designs, let's take a look at the different subscription models and how to get the hobbyist (free) version.

Various subscription options

There are a few different pricing options regarding Fusion 360 and knowing which one to get can save you a lot of money and wasted time. If you are a student or hobbyist, you can use it for free; if you are a professional, then you can purchase a monthly or yearly subscription, as well as get various other add-ons.

The free hobbyist version

This version is the most common for people using Fusion 360 and is what helped Fusion 360 get off the ground as a useful tool for hobbyists and students alike. Back in 2020, Autodesk changed what the free version used to offer, such as the free use of multiple simulation tools, unlimited part design storage space, and limitations on some tools in other workspaces. The new hobbyist version can also be a bit confusing regarding how to access it as the free 30-day trial version is the only free version that they advertise on the Autodesk website. To access the free hobbyist version, follow these steps:

1. Go to the Autodesk Fusion 360 website: `Autodesk.com/Products/Fusion-360`.

2. Click on the **Download free trial** button.

3. There will be two different trial versions on this page, one for business and one for education. The student version will ask for your school student ID. If you do not have a student ID, you will need to get the business version.

4. Once you download and install the business version, your 30-day trial will begin.

5. Once the 30-day trial is up, Fusion 360 will ask you to select a paid version, at which point you can choose the 1-year free hobbyist subscription. You can then renew this version every year.

The top part of your Fusion 360 workspace will look something similar to this when you are using the free for personal use version:

Autodesk Fusion 360 (Personal - Not for Commercial Use)

Figure 1.17 – The top left of the screen showing the personal us license

At the top right of your screen, you will notice that you have a limited number of projects that can be active at one time. This doesn't mean that you are limited to 10 only, just that you can only have 10 active at once:

Figure 1.18 – The top right of the screen showing the limited
number of projects of the personal use license

To be able to switch from an active project to an inactive project or vice versa, do the following:

1. Click on the icon for the limited number of projects; this will open your data panel.

2. Select any project within the data panel that you would like to activate or deactivate.

3. Your number of selected projects will then change, depending on your selection.

The paid version

This version gives you unlimited cloud storage space for your projects, no limitations on creating drawings, and limited access to simulations. To get full access to simulations, you will need to purchase an extension.

Extensions

Once you have a paid version, you can add any extra extensions that may help your design, such as simulations, generative design, advanced machining, and fabrication. Each has its own pricing and learning curve. We will not go over this in this book, but if you become more advanced, these extensions will greatly help your designs take on a new life.

To find out more about the changes within the different versions, go to `Autodesk.com/Campaigns/Fusion-360-Personal-Use-Changes`.

Summary

In this chapter, you learned how to navigate the Fusion 360 UI by learning where the help menu is located and how to use the **Learning Panel** area. You learned where your project files are stored within the data panel, how to change to the seven different workspaces and understood what they do, and learned how to manipulate your toolbar to add buttons that you may use more frequently. We also looked at the different design approaches, such as top-down and bottom-up, and how merging the two will make you a stronger designer. Finally, you were introduced to the different subscription options of Fusion 360. At this point, you should be able to decide which one will work best for you.

In the next chapter, we will learn about planes, sketches, constraints, and parametric constraints.

Planes, Sketches, Constraints, and Parametric Dimensions

In this chapter, we will go over the most typical tools that make Fusion 360 powerful to use as a design program. I'm going to take my time to explain how each of these tools works throughout this chapter so that when you are working on your projects, you will have a better understanding of what they are doing and how they work. By the end of this chapter, you will have a better understanding of these powerful commands and a better idea of what they can and can't do for your project.

In this chapter, we're going to cover the following main topics:

- What are planes?
- What are sketches?
- What are constraints?
- What are parametric dimensions?

Technical requirements

Be sure to have Fusion 360 downloaded and installed on your machine. The interface will look the same for Mac or PC, so you won't have to worry about any difference in icons.

What are planes?

Planes are flat surfaces that you can use to create sketches, use as references to build other geometry, or use as cutting tools. When you first start Fusion 360, you will notice that you have a large grid with red and green axes and a central dot. The dot is your origin (0, 0) location, with the red line being your x axis and the green line being your y axis:

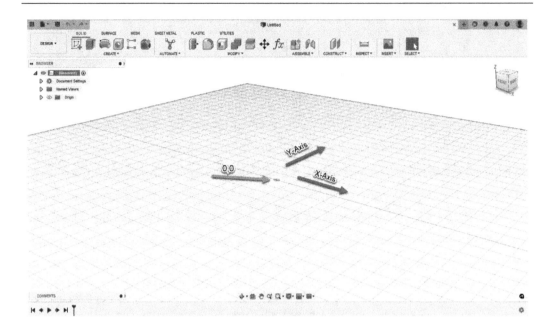

Figure 2.1 – Origin location on a grid

Choosing any tool in the toolbar, such as the **Box** tool under the **SOLID** tab within the **CREATE** panel, you will see three planes appear on your screen with a **Select a plane or planar face** option:

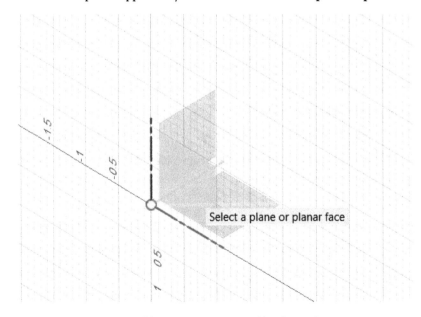

Figure 2.2 – Planes once a command has been chosen

Think of these planes as your drawing paper, and you can choose which side of an object you want to draw on. These planes represent the top, side, and front views of your object.

Once you choose a plane by left-clicking on it, you can then draw the size of the object onto that plane. Try this yourself, as follows:

1. Left-click on the **CREATE** panel drop-down arrow located within the **SOLID** tab and then left-click on the **Box** tool, which is located about halfway down the menu. As soon as you pick the **Box** tool, you will see the planes appear.

2. Now, choose a plane—for instance, the top plane (any plane will work though).

3. Click and drag out a box; the size doesn't matter.

4. Once you choose the size, Fusion will then ask you for the height. You can either left-click and choose a height value or type one in:

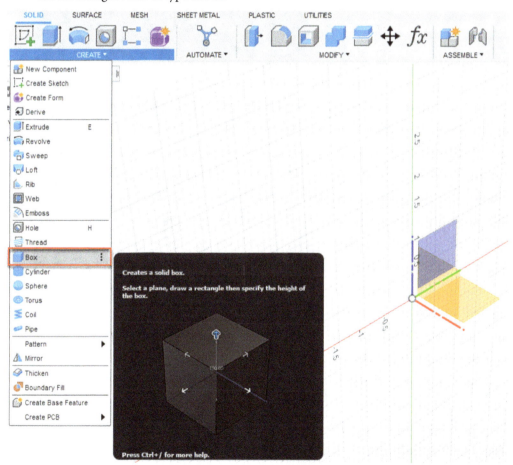

Figure 2.3 – The Box tool and the axes planes appearing

There are a variety of different plane tools in Fusion 360 that you can use to create your objects. These are all located within the **CONSTRUCT** panel dropdown. We will go over only a few of them now as these are the most commonly used.

Offset Plane

One of the most commonly used plane tools is the **Offset Plane** tool. It allows you to offset a flat, planar surface to a specific distance so that you can create another 2D sketch or project 2D sketch objects onto it. You can use this plane tool any flat surface; curved surfaces will not work. This plane tool can be used with the **Slice** tool in order to cut a 3D solid at a certain location:

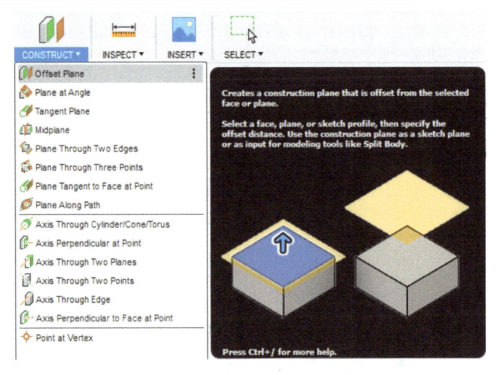

Figure 2.4 – The Offset Plane tool

You can also drag and resize these planes to any size you like after you have created one by mousing over the corner point and clicking and dragging:

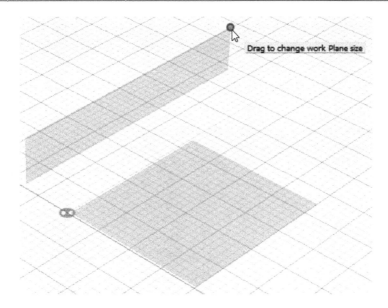

Figure 2.5 – Grip point for stretching planes

The Plane at Angle tool

The ability to create planes at any angle is helpful for creating sketches that may not lay flat on a 2D plane. For example, there may be a drill hole at a specific angle that you need to place on a part. Here's where you can find the **Plane at Angle** tool:

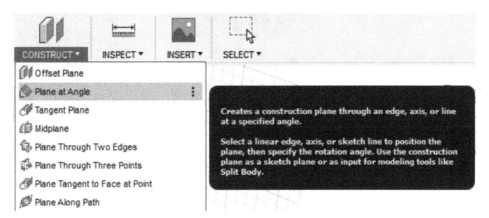

Figure 2.6 – Plane at Angle tool location

You can place planes at an angle on any edge, as shown in the following screenshot:

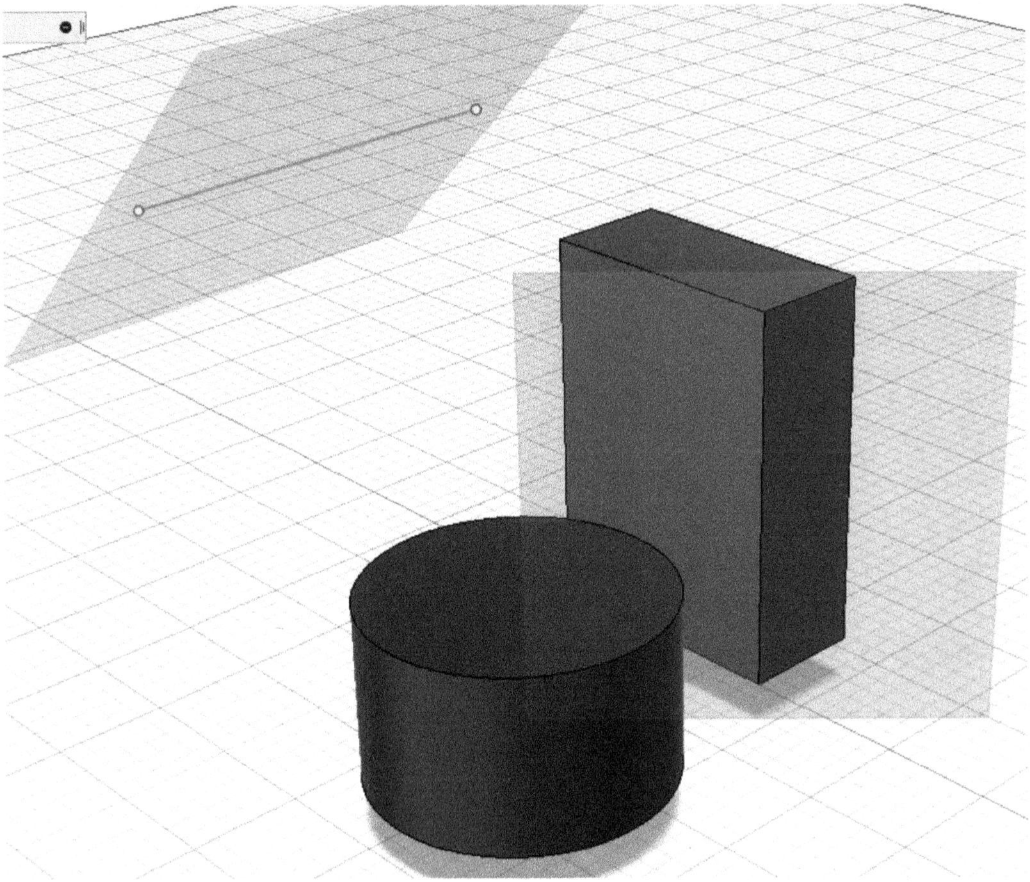

Figure 2.7 – The Plane at Angle tool in use

These edges will not work on circles, but you can use the **Tangent Plane** tool and then specify a degree angle such as 90°, 180°, and so on. Edges also don't need to be on a body since you can also create a line using a sketch; we can use the **Plane at Angle** tool to sketch.

Midplane

Another one of the most used plane tools is the **Midplane** tool. This allows you to place a plane between two faces or other construction planes created. This is helpful when you are trying to create a mirror of another object or when you want to split a body directly in the middle:

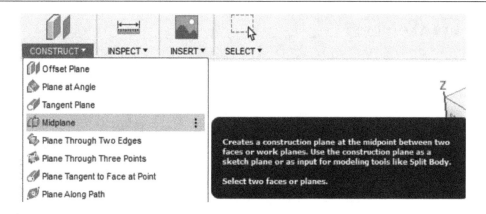

Figure 2.8 – Location of the Midplane tool

Try this tool out yourself, as follows:

1. Choose the **Box** tool, as was done previously.

2. Select **Construct**, then the **Midplane** tool from the toolbar.

3. Pick any two faces of the box that are parallel to each other—that is, left and right, top and bottom, or front and back.

    A plane will appear in between the two chosen faces:

Figure 2.9 – Midplane usage

You will be using planes a lot when using Fusion 360, and there are many different types for different usage purposes. Hopefully, you now have a better understanding of the most commonly used ones and are ready to start placing 2D sketches onto these planes in the next section.

What are sketches?

Sketches are profiles of geometric objects that are built within Fusion 360. These sketches could be just a simple line that creates a surface object or an enclosed circle that generates a solid object. You can create any enclosed shape to generate a solid object, so long as it shows a light blue interior shaded area. Once you have created your sketch and generated your 3D object, you can go back into the sketch to make further changes to your 3D model. You can see the location of the **Create Sketch** tool here:

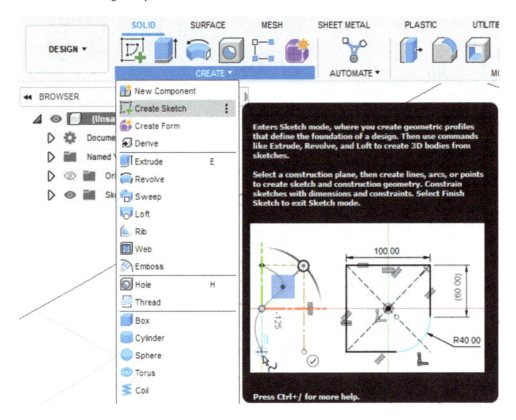

Figure 2.10 – Create Sketch tool location

We will now explore the **SKETCH** environment, which changes the toolbar icons at the top of your screen once a plane is selected and will show new tools for this design space.

The SKETCH tab

Once you have started a sketch by clicking on the **Create Sketch** tool and have selected a plane to draw on, you will notice that your toolbar changes into a new layout:

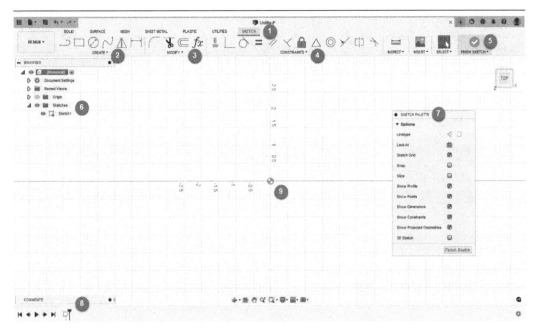

Figure 2.11 – The SKETCH tab toolbar

Let's look at this in a bit more detail:

Number	Name	Description
1	The **SKETCH** tab	This environment allows you to create and modify 2D sketches. If you accidentally click on another tab, click on this button to get back into this environment.
2	The **CREATE** panel	All drawing tools are located here, such as Line, Rectangle, Spline, Mirror, and Dimension. Use the dropdown to see more tools.
3	The **MODIFY** panel	All modification tools are located here, such as Fillet, Trim, Offset, and Change Parameters. Use the dropdown to see more tools.
4	The **CONSTRAINT** panel	Constraints allow you to set rules over sketches, such as Horizontal, Vertical, Parallel, Coincident, and so on. Use the dropdown to see more tools.
5	**FINISH SKETCH**	Click this icon to finish a sketch and go out of the SKETCH tab, which will bring you back to the **SOLID** tab. If you click this icon without drawing anything, the sketch icon will still appear in your drawing history (**8**).
6	The **Sketches** folder	Once you enter the sketch environment for the first time, a new folder will be created for you in your browser. This is to keep your sketches organized. You can double-click on the sketch icon within the folder to edit a sketch.

Number	Name	Description
7	**SKETCH PALETTE**	Contains palette options pertaining to the interface and will show further options of some tools such as rectangle center, construction, circle, and so on.
8	History	Keeps track of the order of operations when objects are created. This can be reorganized to, within reason, allow for model adjustments.
9	Origin point	This point helps to lock down your initial location of a sketch. It also serves as a 0,0 point for the drawing grid.

3D sketches

When you activate the **3D Sketch** tool within the **SKETCH PALETTE** options, you will be able to draw off of your chosen sketch plane:

Figure 2.12 – 3D Sketch option within the SKETCH PALETTE

options and a 3D drawn line in perspective view

In order to see this tool more clearly, you can either use your three-button mouse and hold down the *Shift* + the middle mouse button and move your mouse or you can left-click on the **View Cube** in the top-right corner of your screen and rotate it to a perspective view by left-clicking and dragging on the cube body or by selecting a corner point of the cube. You can now draw off of the grid in any direction. Keep lines straight by staying on an axis such as x (red), y (green), or z (blue), and as you move, you will notice that a color matching the color of the axis will light up.

One reason to use the **3D Sketch** tool would be to create piping with the **Pipe** tool, which we will discuss in a later chapter. You can see the location of this tool here:

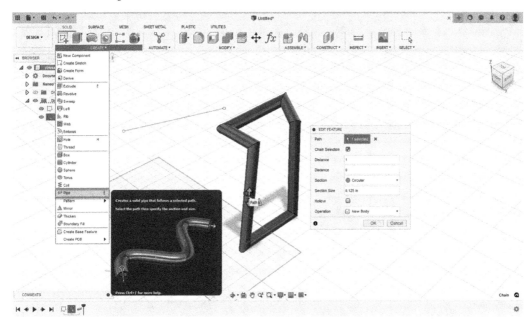

Figure 2.13 – The Pipe tool used on a 3D sketch

Just drawing 2D or 3D line geometry is not enough to create nice, workable 3D objects in Fusion 360. You will need to add dimensions and constraints in order to get the most out of the platform. The next section will explain how to set rules, also known as constraints, for 2D geometry.

What are constraints?

There are multiple ways to set rules to geometry within Fusion 360. These rules that you set allow the geometry to *flex* only in ways that you would like it to move, such as always making a line stay horizontal or two lines parallel. Fusion 360 will automatically add constraints to 2D objects while you draw depending on how closely it matches a constraint. In order to see the **CONSTRAINTS** panel, you will need to have a sketch active and be within the **SKETCH** tab:

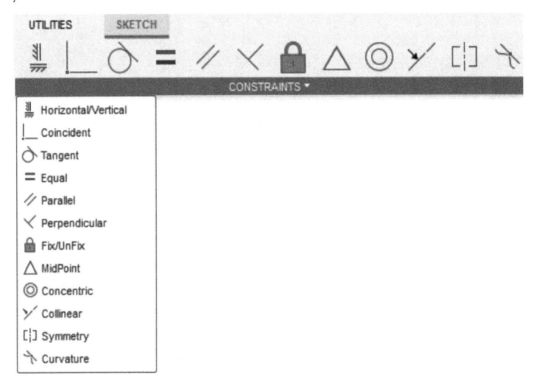

Figure 2.14 – The CONSTRAINTS panel within the SKETCH tab

In the next few sections, we will go over how to add constraints manually, what they do, and also how to remove them when you need more control over your 2D sketch.

Placing constraints nonautomatically

To place constraints without the assistance of Fusion 360, you will need to have a 2D object placed on the screen without a constraint already added to it. You have to then click on the constraint that you'd like to add and then click on the 2D object or multiple objects, depending on the type of constraint chosen. You can see an example in the following screenshot:

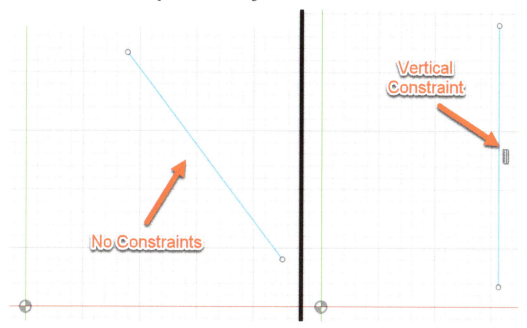

Figure 2.15 – Showing an unconstrained and a constrained 2D line

You'll discover that you'll use some constraints more than others when working on designs, so here are two of the most common (we will go over more as we work):

Icon	Name	Description	Usage
	Horizontal/Vertical	Sets lines or two points to be limited to within the x or y axis	Draw a line at an unknown angle, select the Horizontal/Vertical tool, and use the pick box to select the line. Depending on the slope, the line will be either vertical or horizontal. An example is shown next:

Icon	Name	Description	Usage
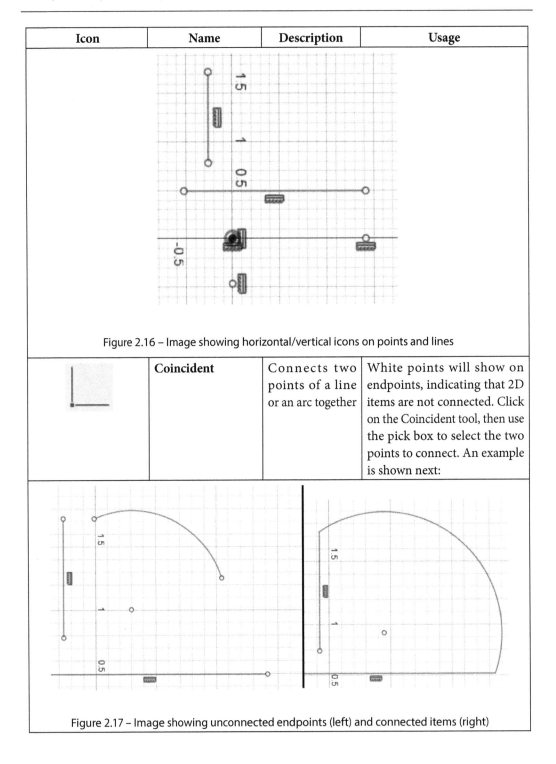			

Figure 2.16 – Image showing horizontal/vertical icons on points and lines

| | Coincident | Connects two points of a line or an arc together | White points will show on endpoints, indicating that 2D items are not connected. Click on the Coincident tool, then use the pick box to select the two points to connect. An example is shown next: |

Figure 2.17 – Image showing unconnected endpoints (left) and connected items (right)

Removing and adjusting constraints

To remove any constraints that are set automatically by Fusion 360, select the icon on the object to which the constraints pertain, then hit the *Delete* key on your keyboard. You can then manually select the constraint that you wish to place:

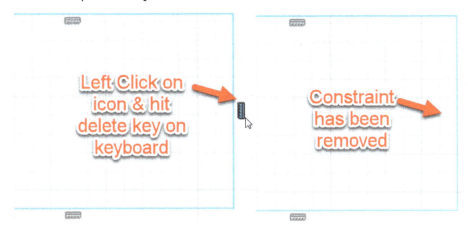

Figure 2.18 – Removal of a constraint

If a constraint has no dimension attached to it, the 2D sketch object can be dragged around the screen by left-clicking and holding the endpoint, line edge, or center point of an object.

What are parametric dimensions?

After you have set up the rules for your 2D sketch, you will need to constrain it further with parametric dimensions. The tool to do this is located under the **CREATE | Sketch Dimension** option (as shown in *Figure 2.19*) or you can just use the *D* shortcut key on your keyboard.

Parametric dimensions allow you to control your 2D sketch with specific dimensions that are editable at any time by either double-clicking on the dimension text or by going into the **Change Parameters** tool:

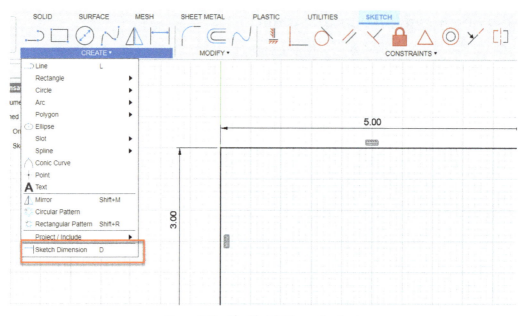

Figure 2.19 – The Sketch Dimension tool

Fully constraining your sketch

When you constrain or lock your sketch, your 2D sketch drawing will be fully locked down, and lines, arcs, and circles will all show as black instead of blue:

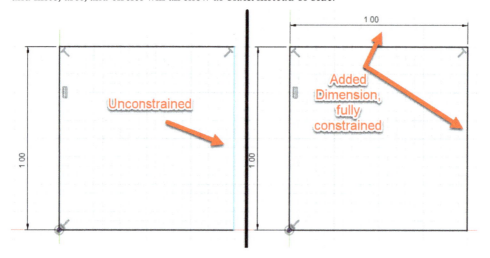

Figure 2.20 – Showing an unconstrained line; the black lines are constrained

A fully constrained sketch will only show fully black when all objects have constraints and dimensions allied to them. Be aware that you may have all the constraints added and all you may need to include is a dimension to finish fully defining the sketch.

Adjusting the size of the dimensions

There are two ways to adjust the size of the added dimensions. The easiest way is to double-click on the dimension text. You can also go into the **Change Parameters** tool through **MODIFY | Change Parameters**. The **Change Parameters** tool will list all dimensions that have been added while you are working on your sketch. You can change these dimensions by left-clicking on the number within the **Expression** column. In the **Change Parameters** flyout, you can also add user parameters through which you can give label names to objects, such as **Thickness**, **Height**, **Weight**, and so on; this will give you a better idea of which dimension connects to which 2D object:

Figure 2.21 – The Change Parameters tool location

When you select the **Change Parameters** tool, you will see a dialog as in *Figure 2.21*. By clicking on the **+ User Parameter** button, as shown in *Figure 2.22*, you can add personal names to items, which you can then add to objects in your project. We will go over much more of this while working on our projects throughout the rest of this book:

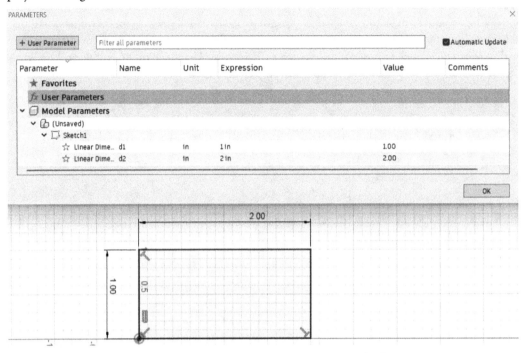

Figure 2.22 – Showing the Change Parameters tool palette

There are many amazing things you can do with the **Change Parameters** tool palette such as adding, subtracting, multiplying, and dividing parameters, as well as performing algebraic functions such as pi, sine, and cosine.

Summary

In this chapter, we learned what planes and sketches are and what constraints can do. These are the basic tools that you'll use daily while using Fusion 360. Knowing about these three basic tools will give you a better understanding of how Fusion 360 works and knowing how you can control these powerful tools will make your learning a much easier process. If you have trouble while working on this book, make sure to come back to this chapter for a quick review and this may help with any problems that may arise while you work. In the next chapter, we will explore setting up 2D sketches and start generating 3D models from those sketches.

3
Project Building Basics

In this chapter, we will learn how to figure out the **design intent** for a project. This means that when you have a design project in mind, the way that it is created will determine how that project behaves when parametric dimensions are added to it. We will explore some ideas and various ways to create objects within Fusion 360 to give you a better idea of how certain parametrics work. By the end of this chapter, you'll be able to understand how design intent works and how to set up a sketch to make it work for you.

In this chapter, we will see the following:

- How to determine design intent
- What Rule #1 is
- Where our files are

Technical requirements

You can practice all projects with the files provided, or feel free to create your own for a more custom experience. The sample design for this chapter can be found at `https://github.com/PacktPublishing/Improving-CAD-Designs-with-Autodesk-Fusion-360/tree/main/Ch03`.

How to determine design intent

When someone asks, "*What is design intent?*", they are asking how a certain object is meant to work and how it will react if changes are made to it. In the following example, you can see two identical models, both with a drilled hole in the middle:

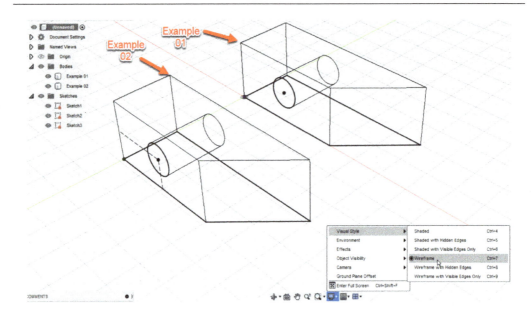

Figure 3.1 – Two models that look the same

Note the location of the hole from the left edge; the hole goes all the way through the solid object. I turned on the **Wireframe** mode via the **Visual Style** dropdown at the bottom of the screen to show the inside of the solid model.

> **Important note**
>
> There are a few different visual styles within Fusion 360 located within the **Display Settings** category at the bottom of the screen. These settings help you customize the way you display your model. Wireframe mode allows you to see within the model, which helps to locate any drilled holes or other hard-to-see features. **Shaded with Visible Edges** is the most common style, since it displays what the model would look like in real time.

Both sketches have the same dimensions and somewhat similar constraints. You can see in *Figure 3.2* that both drawings are located within the same sketch as well, which is possible within Fusion 360; however, try to keep 3D model objects in their own separate sketch, mainly to help with locating that sketch when changes are needed.

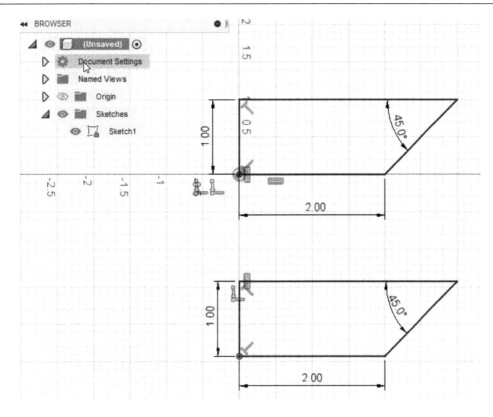

Figure 3.2 – A base sketch that created the 3D models

If we take a closer look, you can see in the first model that the hole has been placed with dimensions:

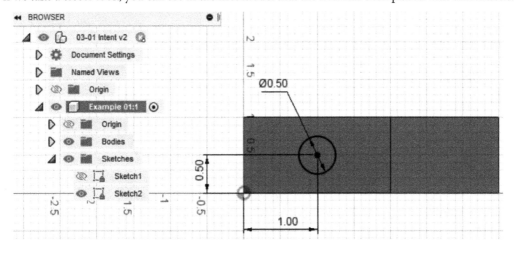

Figure 3.3 – An Example 01 set with dimensions

In the second example, you can see that the hole has been placed with constraints using the midpoint of the left and bottom edges:

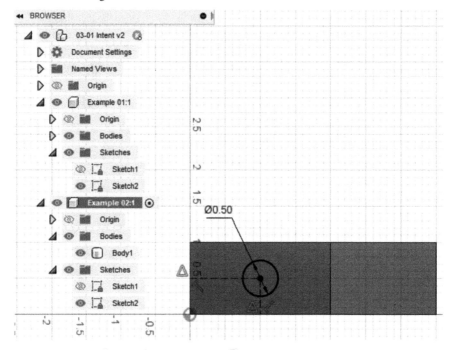

Figure 3.4 – An Example 02 set with constraints

If I were to adjust the dimensions in both examples (see *Figure 3.5*) to a larger size, then the model would "flex" according to the parameters that we added.

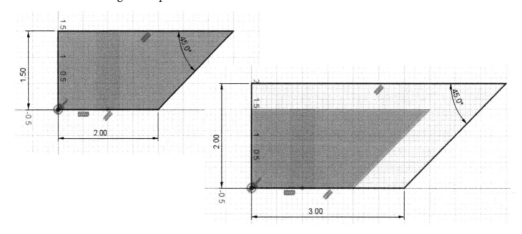

Figure 3.5 – Changed dimensions

You can see that in example 1, the hole has stayed in the same place due to the dimensions, but in example 2, the hole has moved due to the constraints:

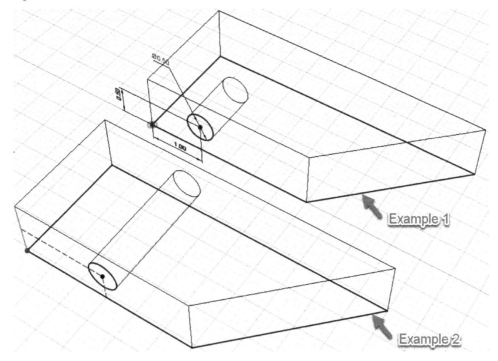

Figure 3.6 – The model has changed size, but the holes differ in location

Both models are created correctly but both have different intents. In the first example, I wanted the hole to stay exactly in its location near the left side edge, so I gave it dimensions with non-changing values.

In the second example, I wanted the hole to adjust with the size of the model and always stay in the center, so I didn't use any dimensions and let the constraints move the hole when the sketch is updated. I also added a rule to the hole length so that when the size changed in **Example 1**, the hole would stay the same length but in **Example 2**, the hole will change with the model.

If you'd like to test this yourself, you can either download the *Chapter 3* example file from GitHub or follow along with these instructions:

1. Start a new sketch by clicking on the **Create Sketch** button.

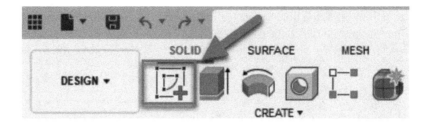

Figure 3.7 – The Create Sketch tool location located within the Create panel

2. Left-click on the *XY* planar face:

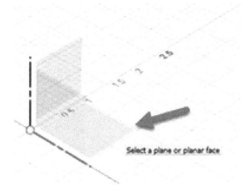

Figure 3.8 – The XY planar face

3. This will now bring you into the sketch environment (notice that the ribbon has changed). Click on the Line tool.

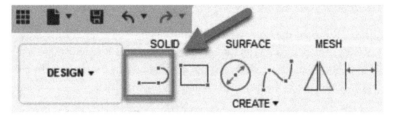

Figure 3.9 – The Line tool location within the Create panel

4. Left-click on the origin dot, and draw a shape similar to the diagram in *Figure 3.10* by left-clicking; don't worry about dimensions or the shape just yet. If you accidentally typed in a dimension, just hit *Delete* to remove it.

Figure 3.10 – The origin dot location

5. After you have drawn the basic shape, add constraints (similar to the following figure) by clicking on the matching icon in the **CONSTRAINTS** panel. The one on the far left is the horizontal/vertical constraint, and the other constraint is the parallel constraint.

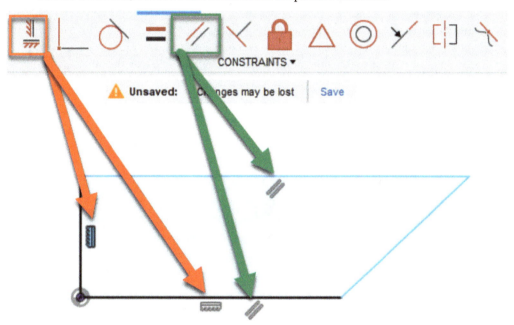

Figure 3.11 – A simple, basic example shape showing constraints

Note that some lines are black and some are blue. This is due to the bottom-left line being connected to the origin. If it was not created on the origin, all the lines would be blue, meaning that they are not constrained or locked to anything.

6. After the constraints, add in the dimensions by pressing the *D* shortcut on your keyboard. You will see that your cursor will show a small dimension next to it, which means the Sketch Dimension tool is active.

Figure 3.12 – The Sketch Dimension tool

7. Left-click once on the edge by the origin first, and then move the mouse left away from the line. You may have a different numbered dimension showing than mine, but that is OK; we will change it in a moment. Left-click to place the dimension in its location, and then hit *Enter*. Repeat this process for the other dimensions by clicking on the line and moving your mouse slightly away from the line, then left-click once more to place it, and hit *Enter*. For the angled dimension, click on one of the lines of the angle, and then left-click on the other.

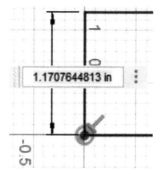

Figure 3.13 – Placing a dimension

8. Once you have all the dimensions placed, your sketch lines will all show black. Now, double-click on the text of the dimensions and it will highlight, allowing you to change it. Change the dimension text to match *Figure 3.2*. If you accidentally placed a dimension and need to get rid of it, first close the Sketch Dimension tool by hitting *Esc*, then select the dimension, and hit the *Delete* key on your keyboard.

9. Now that you've changed the dimensions, click on the **FINISH SKETCH** button at the top right of the panel list.

Figure 3.14 – The FINISH SKETCH button

10. You will now be sent back to the Fusion Design workspace, and your sketch will be shown on the plane that you first chose to draw on. Now, click on the extrude button.

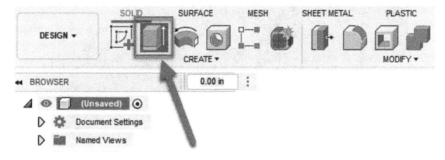

Figure 3.15 – The extrude button

11. Fusion will automatically choose the sketch that you just drew; if not, left-click on the sketch, and it will become highlighted in blue:

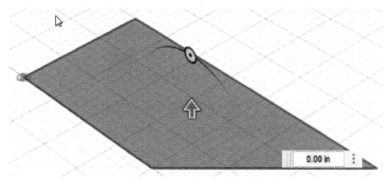

Figure 3.16 – Selecting the sketch profile

12. Type in an extrusion height of 1, and you will now have a solid model. If the solid model did not appear and you do not see a highlighted blue object to pick from, you may not have all the sides connected. You will need to go back to your sketch and check your endpoints. To do this, you can double-click on the sketch icon in your history bar at the bottom, or you can right-click on the sketch and choose **Edit Sketch**.

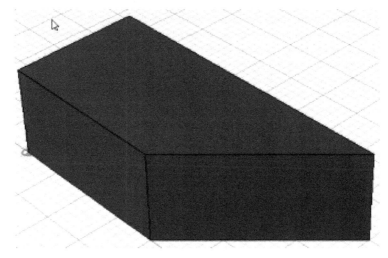

Figure 3.17 – An extruded solid model

13. To place a hole on the side, left-click on the face and click on the **Sketch** button. Fusion will bring you back to the sketch environment. Click on the Circle tool.

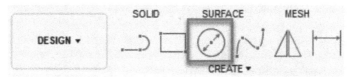

Figure 3.18 – The Circle tool location within the Create panel

14. Left-click and place the circle in a general location on the side of the face; the location and size don't matter, as we will be adding dimensions to the circle.

Figure 3.19 – Place the circle around this area for now

15. As before, click on the Sketch Dimension tool or press *D* on your keyboard. Left-click on the circle, and then left-click on an edge; note that you can only add a dimension to the center of the circle. Place the dimension by left-clicking, and then hit *Enter*. Repeat for the opposite edge. You'll see that the circle is still light blue, meaning that it is not fully constrained. Hit *D* on your keyboard and left-click on the circle itself, then move your cursor away, left-click once more, and hit *Enter*. Everything should be constrained and show as black. Adjust the dimensions to be similar to the following figure.

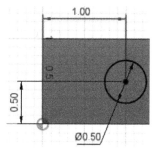

Figure 3.20 – Placing dimensions for the circle

16. Click the **FINISH SKETCH** button to exit the sketch.

17. Once you are back in the design environment, click on the extrude tool and select the circle, if not already selected.

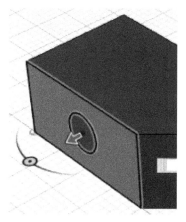

Figure 3.21 – Selecting the circle profile to extrude

18. Left-click and drag the arrow all the way through the object, and hit *Enter*.

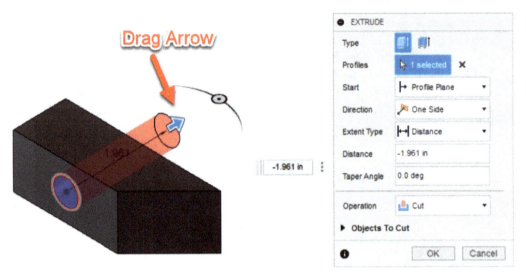

Figure 3.22 – Extruding and cutting beyond the solid body

19. Go over to the **EXTRUDE** window, click on the **Extent Type** dropdown, and select **All**. This will ensure that the dragged-out arrow only goes through all of the object, and nothing else.

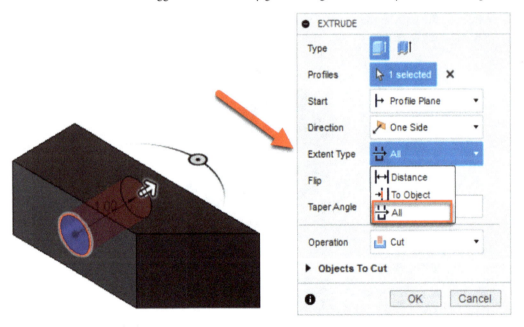

Figure 3.23 – Setting the extent type to All

20. Change the size of the model by changing some of the dimensions, by going back to **Edit Sketch**. Alternatively, you can also edit the dimensions by turning on the sketch name eyeball in the browser, right-clicking on **Show Dimensions**, and double-clicking on the dimension text.

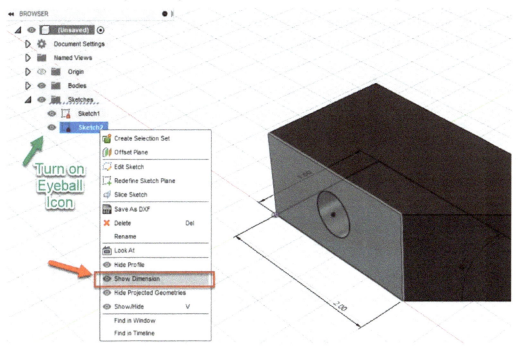

Figure 3.24 – Right-click on a sketch name and select Show Dimensions

Now that you have a taste of what Fusion can do, let's set some rules up so that we know what to do and what not to do.

> **Important note**
>
> When creating extrusions, either by cutting or creating material, pay attention to the extrusion distance number on the right side of your screen. This number may have a minus sign in front of it that tells you which way your extrusion travels. This can throw off new students who learn extrusion, since the direction may go in the opposite direction than what was intended. If that happens, either remove the negative sign or add one in front of the distance number to achieve the extruded direction you want to achieve.

What Rule #1 is

When someone using Fusion 360 mentions **Rule #1**, they refer to an unwritten rule within the Fusion community that, when first starting a design, you need to be sure to use **components**. If you don't plan on using components, your timeline and browser will get very large, making it difficult to locate objects. This helps not only you but also anyone else who you may want to collaborate with on a project.

What components are

A component is basically like a top-level assembly of a part. You can also think of it as a large box that stores all the basic pieces that make up your 3D part, which could be the sketches, the construction planes, the solid body, and so on. This helps with the organization of parts and will shorten your timeline at the bottom of the Fusion 360 user interface if you have a large assembly model.

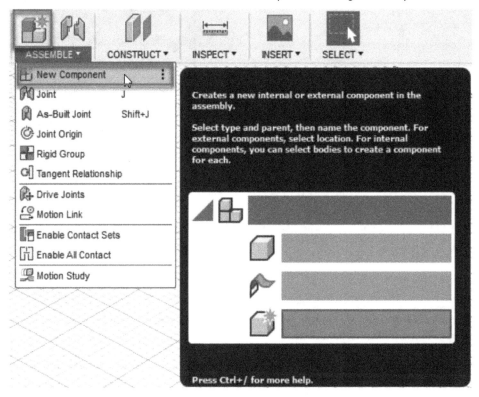

Figure 3.25 – The location of the component tool

If you look at this project example of a tool shed, you can see that all its parts are located within components. This serves multiple purposes, such as organization within the browser and timeline; in addition, components are used to create a **bill of materials (BOM)**.

Figure 3.26 – Components within the browser

> **Important tip**
>
> A BOM is a list of items placed on a drawing sheet that are used to order the parts to build a product assembly. These items could be electrical equipment, wood materials, or mechanical parts.

As you can see in the following figure, the BOM has four items and a quantity that varies, depending on the number of times an item is used in the project. If I left everything as bodies or surfaces, I would not be able to generate anything for the BOM, since Fusion 360 only pulls BOM information from components.

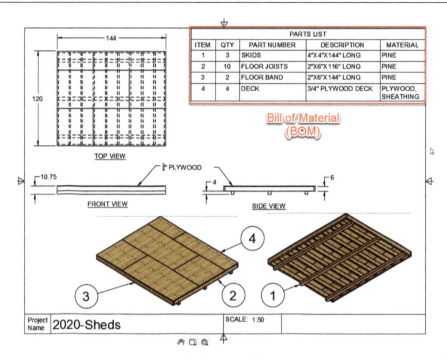

Figure 3.27 – A BOM list created from components

If you happen to forget to create a component when you first start a new part, you can still select a body in the browser – right-click on the body name and then select **Create Components from Bodies**.

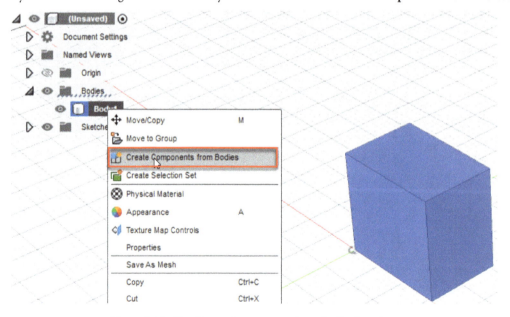

Figure 3.28 – The Create Components from Bodies location

What bodies are

When you first start Fusion 360 and generate your first 3D model from a sketch, that model is usually a body. The only way that you would not generate a solid body is if your sketch was not fully enclosed, which, if you extruded it, would generate a surface body, which can also be placed inside a component. The major difference between a solid body and a surface body is that a solid body can be used to pull material weight information, such as its mass, due to it being a solid object. A surface body has no mass and, therefore, cannot pull mass information until it is turned into a solid body.

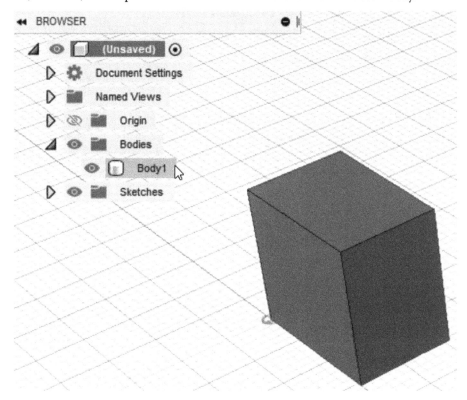

Figure 3.29 – A solid body located within the Bodies folder

The difference between a component and body

Many new Fusion 360 users can get slightly confused by the difference between a component and a body and when to use one or the other. Here is a good example to show the difference between the two. If you take a closer look at the Shed project browser, you can see that it contains a variety of items. Also, note that many of these items are named according to the part that they resemble in 3D, which is a good habit to get into while working on a Fusion project.

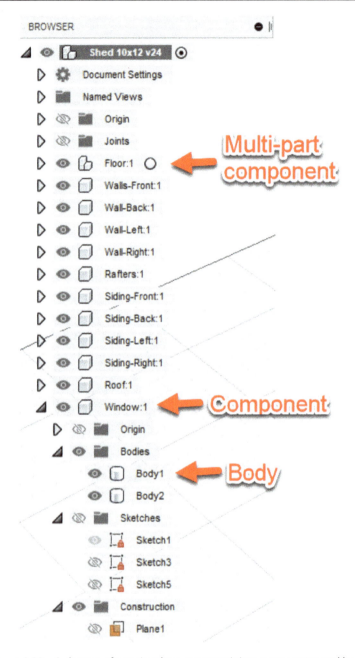

Figure 3.30 – A closeup of a project browser containing components and bodies

As you can see from the preceding screenshot, a component is something like a container or box. This container can sometimes contain a sketch, construction plane, body, and the origin of the individual parts. You can even have components within components, creating a multi-part component.

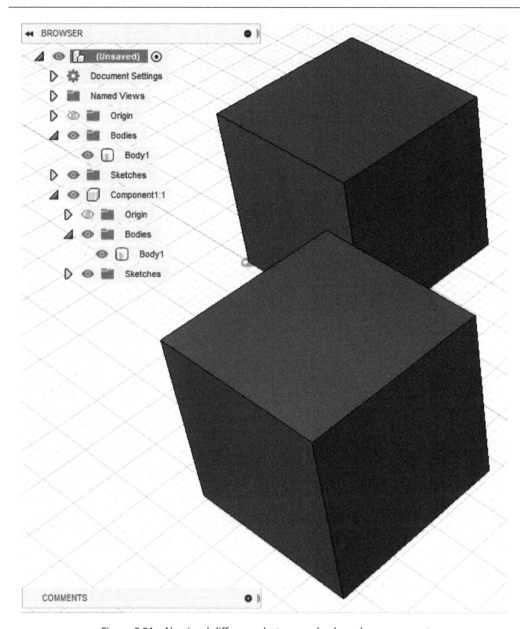

Figure 3.31 – No visual difference between a body and a component

On screen, visually there are no differences between a body and a component; just remember that a component is merely a container that can hold items and is used to build a BOM.

Now that you know how to determine design intent and what Rule #1 is, we can explore where our design files are stored and how to pull them from the cloud.

Saving to the cloud

For those who may not be aware of what the term "the cloud" means, it refers to data storage within a server that may or may not be local to your area. The cloud is useful, as it allows you to be able to work from any machine without having to worry about bringing your personal laptop, or remembering to download your files to a USB stick or external hard drive.

You may have noticed that when you save your project, by hitting the **Save** button in the top-left corner of your screen, there is no option to store your project on the local hard drive. It is saved to the cloud and stored there. You can still create multiple project folders as you would while storing them on your local drive, and you can also create multiple folders within.

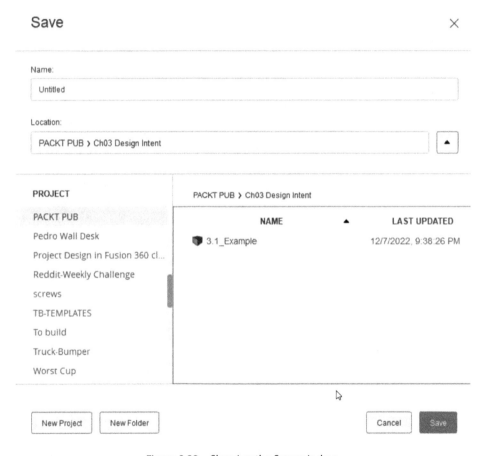

Figure 3.32 – Showing the Save window

If you have the Fusion 360 for personal use subscription, you will only be able to work on 10 files at a time, but this isn't the limit. You can archive some older projects so that you can continue to work on others.

How do I download models from Fusion 360?

You still have the ability to download your projects from the cloud and onto your local drive. If you go to the top left of your screen and click on the **File** drop-down icon, you will see a list of options. To export a file from Fusion, click on **Export**. Note that if you click on **Save As**, you will only be able to save a project with a different name to the cloud.

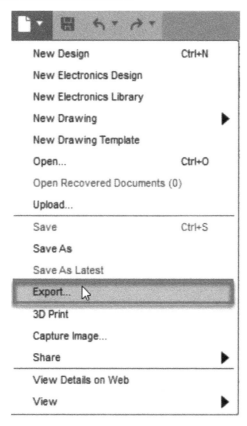

Figure 3.33 – The location of Save As

Once within the **Export** window, you can name the part and click on the ellipsis (**...**) button to save it to a location on your hard drive.

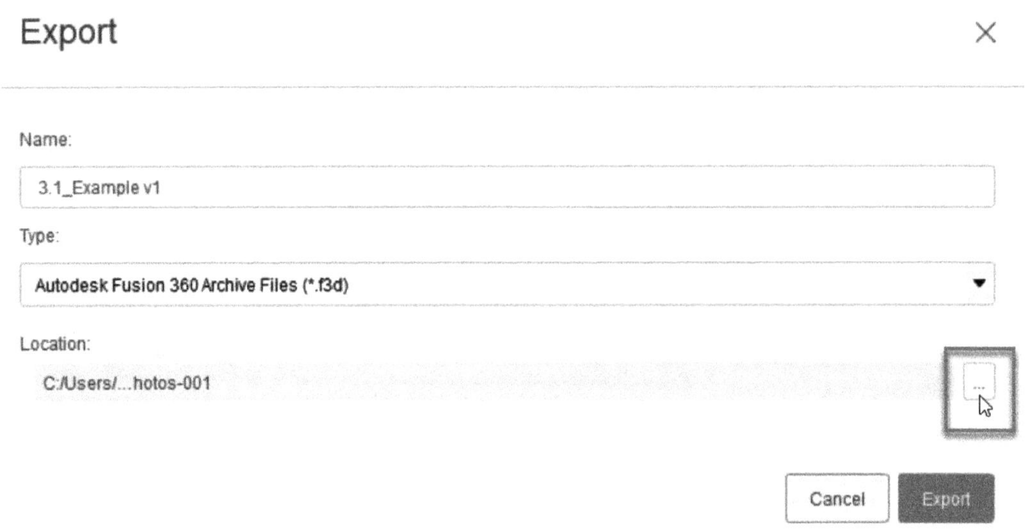

Export   ✕

Name:

3.1_Example v1

Type:

Autodesk Fusion 360 Archive Files (*.f3d)   ▼

Location:

C:/Users/...hotos-001

Cancel   Export

Figure 3.34 – Showing the Location to save button

You have the ability to save your project in a variety of formats, with one of the most common being the STL format for 3D printers.

Figure 3.35 – Showing the list of file extensions for export

Knowing how to export to various file types is helpful when other users work on another software, or if you need to export to 3D printers.

How do I upload models to Fusion 360?

Since Fusion 360 has a cloud storage system, you can store a variety of file types within the cloud. Even if some of these file types may not be able to be imported into the Fusion workspace, they can still be stored to share project information. To upload files into the Fusion cloud space, do the following:

1. Click on the **Data** panel in the top-left corner of the screen.

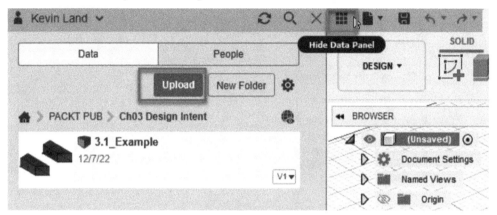

Figure 3.36 – The Upload button within the Data panel

2. Select the project that you are currently working on, create any subfolders to keep an organized file structure, and then click on the **Upload** button.

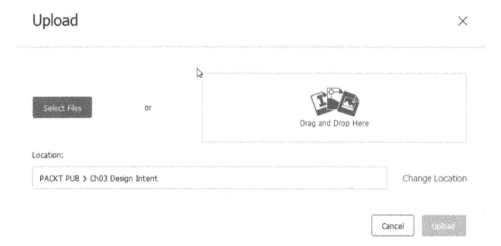

Figure 3.37 – The Upload window

3. Now, navigate to the folder that contains the files that you want to upload. In the following list (*Figure 3.38*), you can see a list of the various file types that can be used in Fusion 360.

```
All Files (*.*)
Alias Files (*.wire)
AutoCAD DWG Files (*.dwg)
Autodesk Eagle Files (*.sch *.brd *.lbr)
Autodesk Fusion 360 Archive Files (*.f3d *.f3z *.fsch *.fbrd *.flbr *.f2t)
Autodesk Inventor Files (*.iam *.ipt)
CATIA V5 Files (*.CATProduct *.CATPart)
DXF Files (*.dxf)
FBX Files (*.fbx)
IGES Files (*.ige *.iges *.igs)
NX Files (*.prt)
OBJ Files (*.obj)
Parasolid Binary Files (*.x_b)
Parasolid Text Files (*.x_t)
Pro/ENGINEER and Creo Parametric Files (*.asm* *.prt*)
Pro/ENGINEER Granite Files (*.g)
Pro/ENGINEER Neutral Files (*.neu*)
Rhino Files (*.3dm)
SAT/SMT Files (*.sab *.sat *.smb *.smt)
SolidWorks Files (*.prt *.asm *.sldprt *.sldasm)
SolidEdge Files (*.par *.asm *.psm)
STEP Files (*.ste *.step *.stp)
STL Files (*.stl)
3MF Files (*.3mf)
SketchUp Files (*.skp)
123D Files (*.123dx)
```

Figure 3.38 – A list of import file types

You can also upload PDFs and image files such as JPGs, PNGs, and TIFFs, which can be used as reference background images.

How do I share my project with others?

If you want to show others the project that you are working on, you can collaborate and work together with them. If you click on the **Data** panel in the top-left corner of the screen, it will open, allowing you to navigate to a project. Click on a project that you are working on, and toward the top of the screen, you will see two buttons, one labeled **Data** and another **People**. Click on **People**, enter a collaborator's email address, and then click **Invite**. This will send an invite to their email, which will prompt them to create a free Autodesk account. Once they create an account, they will be able to look at the project that you shared. Be aware though that this is shared at the project level. You can't share just one folder. All folders within the project will be shared, not just the one you may be working on.

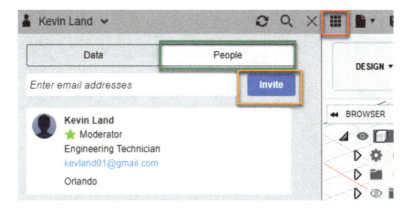

Figure 3.39 – The location of project sharing

Summary

In this chapter, we learned how to specify design intent and think about how a project will be first set up. We learned the difference between components and bodies and the importance of Rule #1. We also learned how to save our files and where Fusion 360 stores them.

In the next chapter, we will start our first project, which is to create a customizable S-hook that, if you have access to a 3D printer, you can print out and use around your household.

4

Creating a Customizable S-Hook

For our first project, we will learn how to create an S-Hook of various sizes using **Fusion 360**. There are a variety of ways to create an S-Hook but, for the sake of time, we will experiment with only two different ways to design an S-Hook using tools such as **Line**, **Arc**, **Sweep**, and **Pipe**. We will learn how to start a project, how to use Rule #1 (see *Chapter 3*), and experiment with different methods of creation.

By the end of this chapter, you will have a better understanding of how to set up a project and get a better understanding of how to fix a project.

In this chapter, we're going to cover the following main topics:

- Project setup in Fusion 360
- Creating an S-Hook using the Pipe and Line tools
- Creating an S-Hook using the Arc and Sweep tools

Technical requirements

In this chapter, you can work with the files provided within the download link or you can build as we go throughout the chapter. Go to this GitHub link to download the *Chapter 4* files: `https://github.com/PacktPublishing/Improving-CAD-Designs-with-Autodesk-Fusion-360/tree/main/Ch04`.

Project setup in Fusion 360

Knowing how to start a project is a crucial step when working in Fusion because if you don't start off on the right foot, it can result in poor design and unexpected results.

It's a good idea to think about and plan out your design before you start drawing anything in Fusion. Let's go over a few thoughts on the design we want to achieve:

- We know the design shape we want, which is an S-shape, but for this S-Hook design, we want it to grow from the center out when the size changes. This way, we can easily control both halves as they grow to different sizes.

- We want the thickness to change when we adjust the overall size.

- We want the ends to have a larger diameter to not allow for slippage.

- We may want to change the shape of the body depending on the purpose of the hook.

Now that we have some idea of the results we want to achieve, let's set up Fusion 360. Open up Fusion 360 and check to see whether any updates need to be installed. You can tell by the clock in the top-right corner of the screen. If there is a red dot on the icon, this means there is an update. Most of the time, it will install automatically unless there is a problem with the internet connection. Otherwise, let it install and then restart Fusion 360 by closing it and opening it back up again. This will make sure that all of the functions work without any installation holdups.

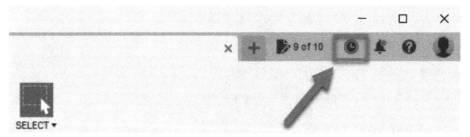

Figure 4.1 – Showing the location of the Fusion 360 updates button

Once all updates have gone through, let's start by saving the design that we are about to work on. Fusion 360 will save automatically for you, but if you didn't save from the very beginning, then there may not be anything to recover in case of an error.

You can save your project by doing the following:

1. Click on the **Save** button; the **Save** pop-up window will open.

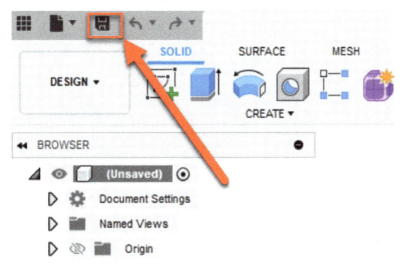

Figure 4.2 – Showing the location of the save button

2. In the **Name** box, type in S-Hook (or anything you like).

3. Click on the drop-down arrow (shown in red in *Figure 4.3*). This will open up the project location area.

4. Click on the **New Project** button (shown in green in *Figure 4.3*). Give the project a name such as Packt Pub or something similar that you will remember.

5. Once you create the project, click on the **New Folder** button (shown in orange in *Figure 4.4*). Name the folder Ch04 S-Hook or something similar.

Figure 4.3 – Showing the Save pop-up window

6.   Finally, click on the blue **Save** button.

It's a good habit to save your design first to avoid losing all your hard work if the program crashes. Now that the project is saved and named, let's start working on the various S-Hook designs.

Creating an S-Hook using the Pipe and Line tools

We will experiment with various ways to create an S-Hook within Fusion 360. There is no one right way to create an object but the design approach you use should depend on your ability to plan ahead and predict how some tools are best utilized and most efficient. As you work in Fusion 360, you will notice various ways to build, and as you experiment with these different ways, you'll notice that some commands work better than others, and while some may generate a 3D object more quickly, they may hinder you in using other tools as you design. Unfortunately, the best way to get over this is to practice and experiment as much as possible. As many months pass, you will soon start to notice how much better your designs are when you can plan ahead and start your projects on the best path.

The quickest and most straightforward way of creating an S-Hook is using the Pipe tool. Using this tool, you first draw a curved line within the **SKETCH** environment. Then, you finish the sketch (see *Chapter 3*) and click on the Pipe tool, which will use a pre-generated shape to create a solid body.

Let's first create a new component so that we can keep our timeline organized since we will make another part within this file:

1. Open the S-Hook file, unless it is already open.

2. Using Rule #1, click on the **New Component** button within the **ASSEMBLE** panel.

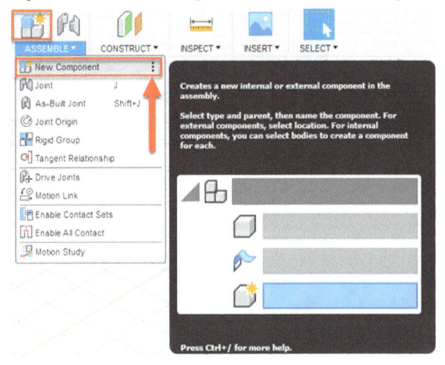

Figure 4.4 – Adding a new component

3. The **NEW COMPONENT** pop-up window will open to the right of your screen. Go to the **Name** field and change the default name from Component1 to S-Hook_Pipe. Then, click the **OK** button.

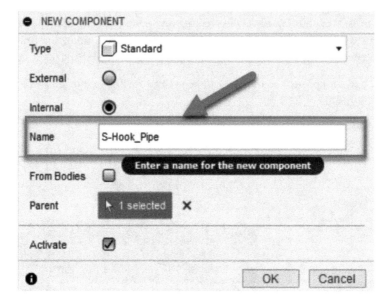

Figure 4.5 – NEW COMPONENT flyout window

This will create a new component and will activate it in the browser. This will happen whenever you create a new component. Beware though that you can create a component within another component, creating a multi-part component, which is good when you want to create one but not good when you want to create a separate component. The best way to avoid this mistake is to take your time and keep a watchful eye on the browser and where that black dot is located.

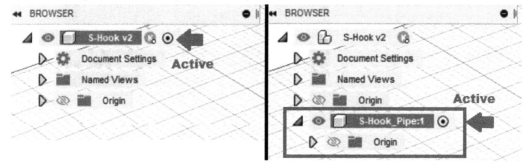

Figure 4.6 – Activating a component

4. Click on the **Create Sketch** button within the **CREATE** panel. (Notice that the **New Component** button is also located within the dropdown. This is typical in most CAD programs where a command is in multiple locations. This is so that users have more options when working.)

Figure 4.7 – The Create Sketch tool location

5. Click on the **XY** plane, also known as the bottom plane. The screen will shift to a view of the selected plane from the top down as if you were drawing on a piece of paper.

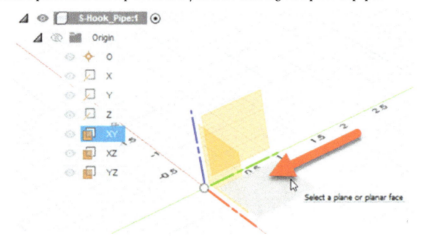

Figure 4.8 – The XY plane location

So far, you have set up the basics of creating a component, which helps to keep your drawing organized. Next, we will start to create the first S-Hook.

Conducting the first experiment

Let's try out our first experiment. I will only go over two different methods, but there are a variety of ways to create a simple S-Hook, such as by drawing a rectangle in the shape of an S and then extruding it to make it 3D. This will create an S with sharp corners all around, which you could keep as is or use the **Fillet** tool to smooth out.

Let's start with an S-Hook that most people would create, so let's draw an S shape using the Line tool:

1. Click on the **Line** tool within the **CREATE** panel and draw a shape similar to the following figure.

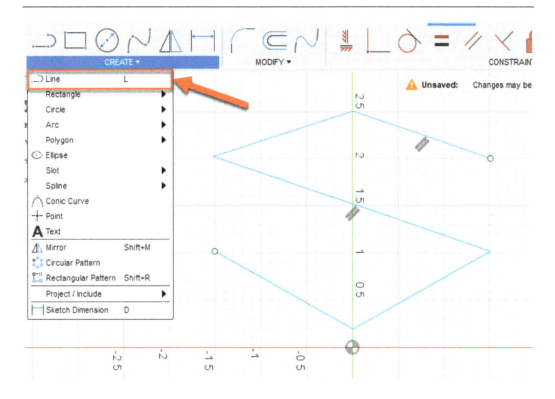

Figure 4.9 – Showing an S-shape created with the Line tool

A few things to note:

- Notice that the line is light blue, which means that this line is not constrained and can move

- If you were to click on the line and drag it onto a point or line edge, it would move to wherever you wanted to place it

- Sometimes, moving is helpful when you are first starting your design, as you can quickly shape your design without worrying about structure

- When you are happy with the shape, you can add in more constraints and then dimensions to fully lock your design down to match your final intent

2. Click on the **FINISH SKETCH** or **Finish Sketch** button. Notice that there are two locations for this operation. Either will work and you can use whichever is faster for you to click on.

Figure 4.10 – Two locations

3. Next, go to the **CREATE** panel and click on **Pipe**.

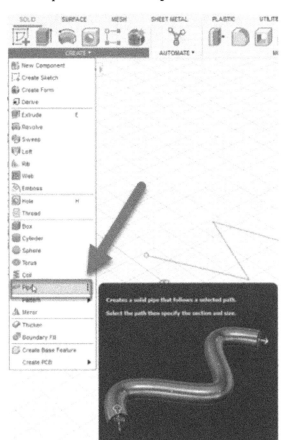

Figure 4.11 – The Pipe tool location

4. Click on the line to select it and you will instantly see the Pipe tool applies default sizes to the pipe.

    The following are some things to note about the Pipe tool:

    - Be sure that the **Chain Selection** box is checked to ensure the tool only selects one line instead of all of them at once.

    - Leave **Distance** at 1 as this will ensure that the Pipe tool continues throughout the line. Feel free to experiment with the **Section** area, as that will change the shape from a circle to a square or a triangle.

    - Experiment with **Section Size** as well and notice that if you change from 0.20 in to 1 in, the thickness of the pipe changes too.

    - Notice that there is a **Hollow** option within the **PIPE** tool window. This option is great for when you need to generate a flexible pipe for your model.

5.   Click on the **OK** button to save the changes.

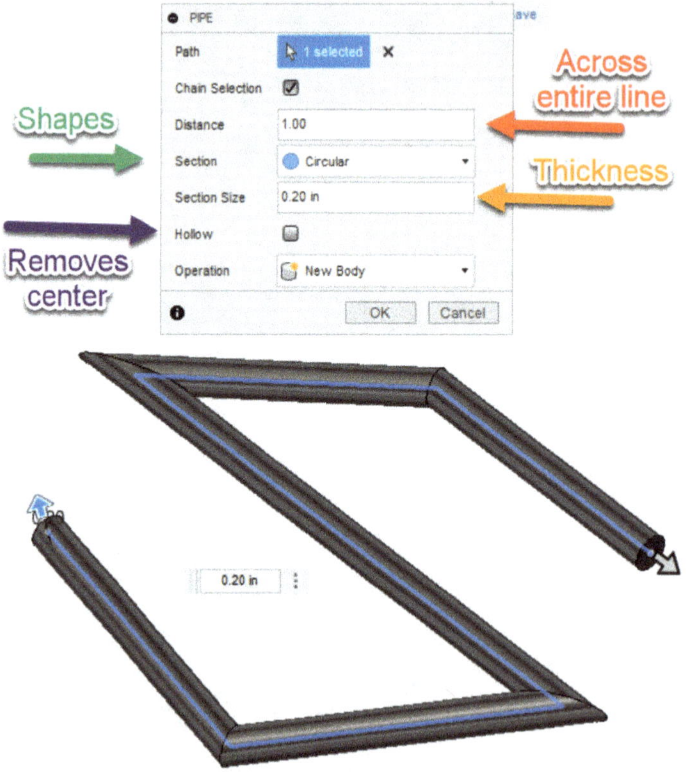

Figure 4.12 – PIPE options flyout for the Pipe tool

Congratulations, you now have an S-Hook... kind of. The corners are sharp, which we could fix with the **Fillet** tool, but if we were to change the size of the S-Hook, we may get unexpected results. Let's open the sketch back up and add a dimension to the lines to change the size and see what happens.

Editing the sketch

To make changes to the existing geometry in Fusion 360, you will need to edit the existing sketch. To do this, we need to go over to the browser on the left-hand side of the screen or edit within the timeline at the bottom of the screen:

1.   Be sure to click **Ok** to finish out the Pipe tool or you won't be able to edit the sketch with a command still active.

2.   Go to **BROWSER** on the left side of your screen and navigate to the sketches folder.

Note the following things:

- The sketch you created will be within this folder but your sketch name may have a different number from the one shown in the example shown in green in *Figure 4.13*

- You have the ability to rename sketches by double-clicking on the name portion, which is a very good idea for large assemblies, as your drawings can get overwhelming very quickly and you may not know which sketch goes to what model

- Keeping your drawings organized will help save time and money

3. To edit the sketch, right-click on the name and choose **Edit Sketch**. You can do this from the **BROWSER** window on the left side of your screen or within the timeline at the bottom of your screen.

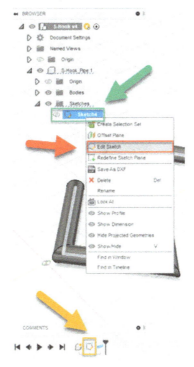

Figure 4.13 – Showing the location of Edit Sketch

4. Once you click the **Edit** icon, you will be brought back into the **SKETCH** environment.

5. Click on the **CREATE** panel dropdown and then click on **Sketch Dimension**. You can also use the keyboard shortcut, *D*, highlighted in green in *Figure 4.14*.

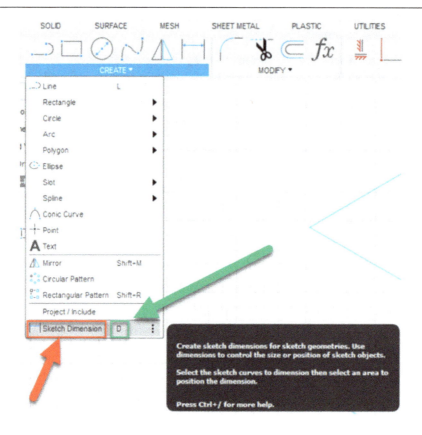

Figure 4.14 – Showing the Sketch Dimension tool

6. Notice that you have a small dimension icon next to your cursor. This indicates that you have the dimension tool activated. If you don't see it, you may have to click on the tool again.

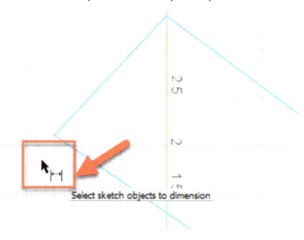

Figure 4.15 – The Sketch Dimension tool next to the cursor

7. With the dimension tool activated, left-click on a line as shown in *Figure 4.16*, and then move your mouse away from the line a little bit. Try to align the dimension as shown in *Figure 4.16* and then left-click once again to place it.

Note the following things:

* As you move your mouse around the screen, the dimension will change from a horizontal dimension to a vertical dimension and then to an aligned dimension. This allows for quick selections of various dimension types.

* Your dimension size may vary depending on the size of the line you created.

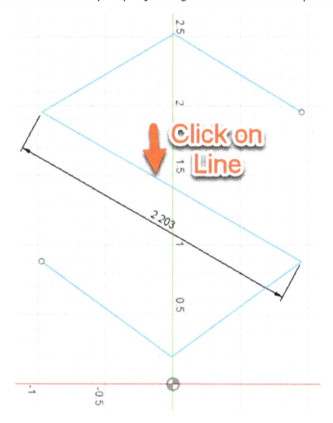

Figure 4.16 – The align dimension

8. Once you have placed the dimension, double-click on the text of the dimension to change it. Let's make it 3 in long.

Note the following things:

- Notice that the line has moved in a direction that you may not have expected

- When lines are not constrained to another point or line, the line will move in any direction

- The reason for this random directional movement is that the sketch was not constrained, meaning an endpoint or midpoint needs to be connected to another point on the sketch

- A lock constraint will also work but typically the origin dot is the preferred location to connect to for the start of a sketch

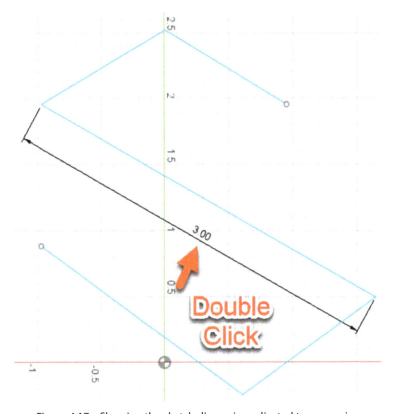

Figure 4.17 – Showing the sketch dimension adjusted to a new size

Let's add some constraints first, which will help to set up rules that the dimensions can follow.

Adding constraints

Adding constraints will help to set up rules that the dimensions can follow. You can either set constraints just by using an icon or you can also use construction lines along with constraints to set rules:

1. Click and drag the bottom point of the S-shape to connect to the origin dot. This sets an anchor point for the sketch so that when a dimension changes size, it knows which way it should move.

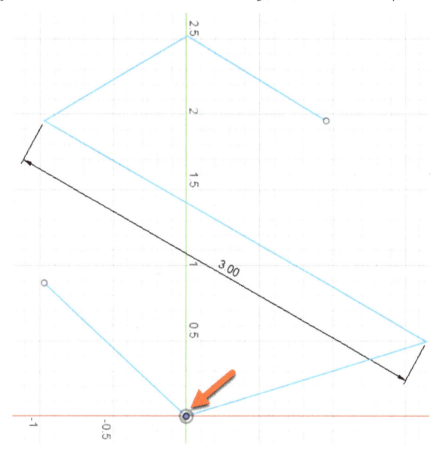

Figure 4.18 – Showing the located endpoint on the origin dot

2.   Click on the **Line** tool within the **CREATE** panel, or you can also use the *L* keyboard shortcut.

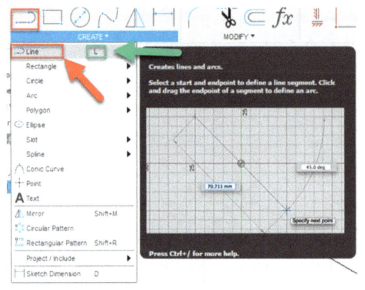

Figure 4.19 – Showing the Line tool location

3.   With the Line tool active, click on the **Construction** button to set the line to a dashed line type. This means that you can use this line but it will not be a selectable line when you exit the **SKETCH** environment.

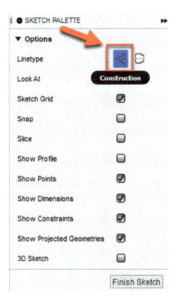

Figure 4.20 – The location of the construction button

4. Left-click on the bottom origin dot and then left-click on the top-center dot as shown by the green arrows in *Figure 4.21*. If the line is not vertical, you can set it to be by clicking on the **Horizontal/Vertical** button highlighted in red in *Figure 4.21* and left-clicking on the line.

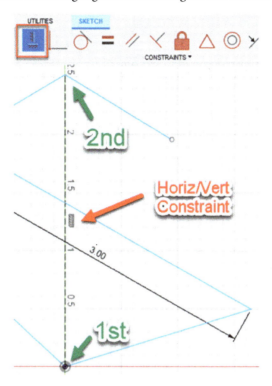

Figure 4.21 – The Horizontal/Vertical constraint and a construction line

5. Do the same for the other two points but set these as horizontal, as shown in *Figure 4.22*. If the dimension gets in the way, you can left-click on it and hit the *delete* key.

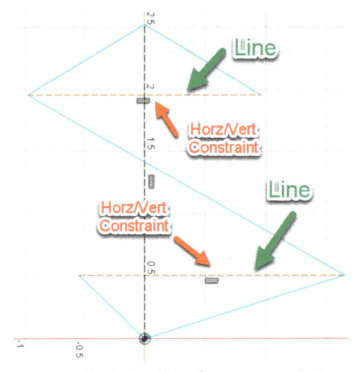

Figure 4.22 – Showing the addition of two more construction lines

6. Use the symmetry constraint to keep the S-Hook symmetric to the center line as shown in *Figure 4.23*.

Take note of the following things regarding the symmetry constraint:

*   The tool works by clicking on one endpoint, then the opposite side, and then the midpoint.
*   Left-click on the symmetric constraint, then left-click on the endpoint labeled **1**, and then click on the endpoint labeled **2** on the opposite side. Finally, click on the symmetric labeled **3**.
*   The endpoints will place themselves symmetrically according to their locations around the middle of the construction line.
*   Do the same for the endpoints on the top half of the grid.
*   Click and drag these endpoints now; the lines will grow and shrink together.

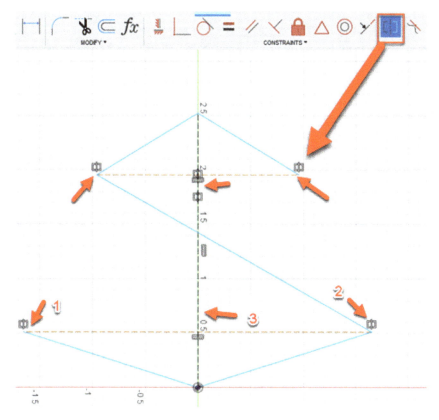

Figure 4.23 – The symmetric constraint location

These constraints should be enough to start adding dimensions.

Adding dimensions

Adding dimensions is the next part of creating a sketch. You normally wait until you have a general shape first, and your constraints, and then add the dimensions last:

1. Click on the **Dimension** tool or hit *D* on your keyboard to use the shortcut key.

2. Add in the following dimensions by left-clicking on a line, then moving your mouse away from the line, and then left-clicking again to place it.

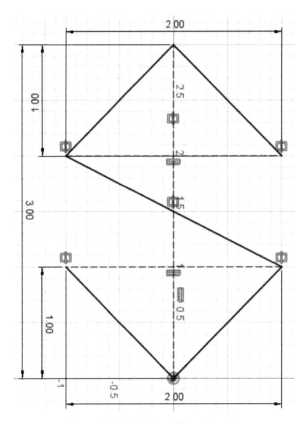

Figure 4.24 – A fully defined sketch

3. Notice that now all the lines are black, meaning that you have a fully constrained sketch. Click on the **FINISH SKETCH** button.

The S-Hook looks good now but it has a lot of sharp points. We can round these off a bit by using fillets. Typically, when working on projects, fillets are saved for last, usually because they may get in the way of your design changes and cause unexpected results. There are two ways to create fillets. You can either add them to the 3D model or to the sketch. Let's explore both ways.

Adding fillets to the 3D model

Let's try adding fillets to the 3D model first. A fillet is added to sharp corners to prevent a person from cutting their fingers on sharp edges. Something to know though before adding fillets is that the radius is limited to the amount of material at that corner point. For instance, if you have a corner point with two lines that are both 1 inch long, you can't add a fillet that is larger than 1 inch; you will get an error:

1.   With the 3D model open in the **DESIGN** workspace, left-click on the **Fillet** tool located within the **MODIFY** panel, or use the *F* keyboard shortcut.

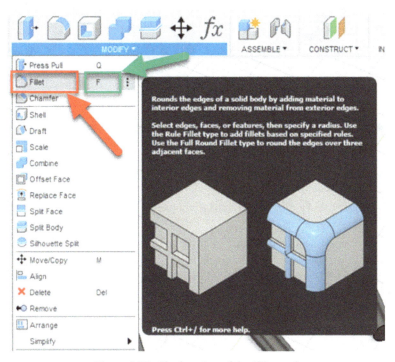

Figure 4.25 – The location of the Fillet tool

2.   The **FILLET** floating panel will appear showing a variety of options. Left-click on the top corner of your S-Hook and notice that Fusion 360 will highlight the edge.

Figure 4.26 – The FILLET tool options panel

3. You can either drag the area indicated by the large blue arrow in *Figure 4.26* or type in the distance within the pop-up box. Notice that if you drag too much, you will get an error due to not having enough material to add to the fillet. Type in a distance of .1 and notice that it creates a small fillet.

You could keep going around adding in fillets but you may not end up with the nice large smooth fillet that you were hoping for. Let's try to add the fillet to the sketch instead.

Adding fillets to the sketch

Adding a fillet to the sketch allows for a bit more ability to change the shape of the sharp edge. Beware though, there are times when it is better to add the fillet into the 3D model since it can be much more easily controlled:

1. Undo any fillets added to the 3D model in the previous section.

2. Double-click on the sketch icon within **BROWSER** or within the timeline below or right-click and choose **Edit Sketch**. Clicking on the eye icon will only turn the sketch on or off, and double-clicking on the name will only allow you to edit the name of the sketch.

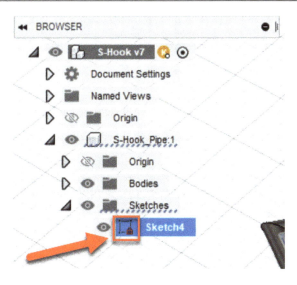

Figure 4.27 – The location of the active sketch

3. Within the **SKETCH** environment, go to the **MODIFY** panel and left-click on the **Fillet** tool, or you can use the shortcut *Shift + F*. Then, left-click on the top corner piece indicated by the red arrow in *Figure 4.28*.

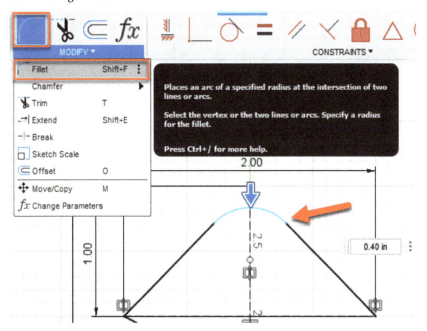

Figure 4.28 – The sketch Fillet tool

The following are some things to note about the **Fillet** tool:

- Notice that by dragging the blue arrow, you can create a much larger fillet

- Type in a value of 1 in and hit *Enter*

- There will be a warning that a dimension has been removed in order to apply the fillet (the warning will fade away, so don't worry if you didn't see it)

- A warning will display when fillets are too large or when constraints are removed in order to ally the fillet

- The warning is shown because when this sketch was created, we used lines that connected at a corner point, and when the fillet was applied, the coincident constraint that attached the two lines together needed to be removed

- You may receive some unknown errors later on due to constraints having been removed

- The best way to get around these types of errors is, with experience using Fusion 360, to know where and how to place constraints

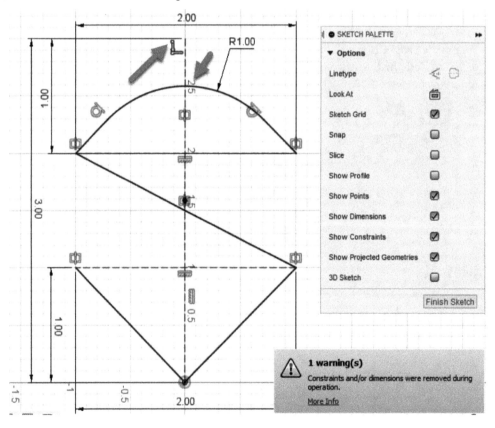

Figure 4.29 – Showing the Fillet tool added and a warning

4. Continue to add fillets to all corners. Notice though that you can't add 1 to all the corners due to the size limitations represented by the orange arrows in *Figure 4.30*.

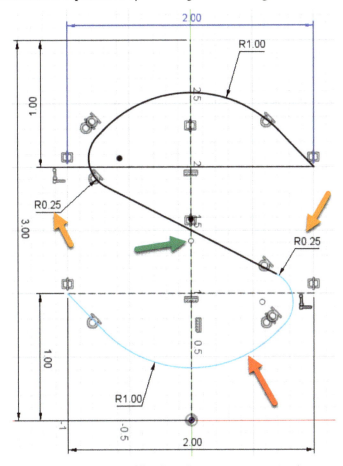

Figure 4.30 – The fillets have become unconstrained

Also notice that when you add 1 to the bottom corner point, you will get the same warning as before and the bottom lines will turn blue, meaning that these have become unconstrained (indicated by the red arrow). This is because the fillet that was added has a center dot that is not connected to anything, as indicated by the green arrow.

5. Left-click and drag the white dot indicated by the green arrow in *Figure 4.30* and notice that it can be moved around.

Here are some things to note about white dots:

- Any white dot you see on your screen can be moved, as it has become unconstrained.

- To repair them, left-click on the white dot and then click on the coincident button in the **CONSTRAINTS** panel. Then, left-click on the center line as indicated by the green arrow in *Figure 4.31*. This will constrain the dot to that line.

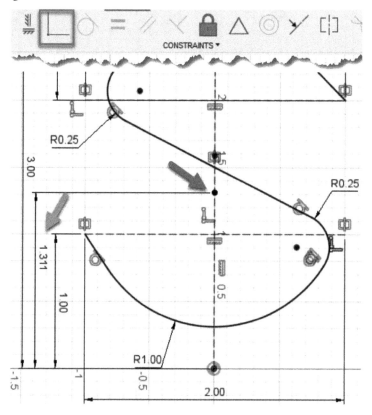

Figure 4.31 – Constraining a white dot to a line

6. Now we need to set that white dot in place by adding a dimension. Use the *D* keyboard shortcut and add a dimension from the origin point at the bottom to the constrained dot. Click on the **FINISH SKETCH** button.

Figure 4.32 – The model that needs rebuilding

You may have to rebuild or reconnect your 3D model since a few changes were made to the sketch. This may happen sometimes while working in Fusion 360. It may not happen all the time but it depends on how much Fusion can remember from the sketch you previously made.

Rebuilding the model

When you clicked on **FINISH SKETCH**, you may have noticed that your model does not look the same. You may have to fix this model using the timeline. Do the following to fix the model:

1. Double-click on the Pipe tool within the timeline or right-click and choose **Edit Feature**.

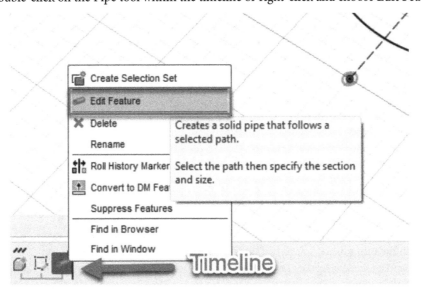

Figure 4.33 – The timeline and Edit Feature

2. Sometimes, you may have to remove the old selection first in order to reapply the selected path. Click on the **X** symbol indicated by the red arrow in *Figure 4.34* within the **Edit Feature** path to remove the old selection.

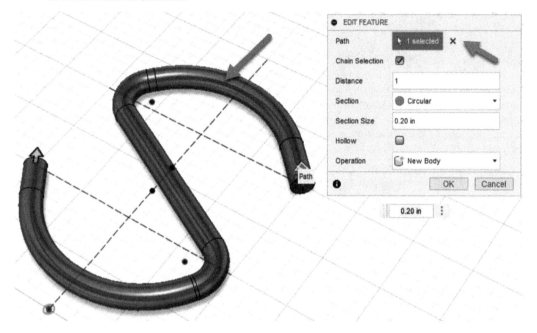

Figure 4.34 – The Path feature

3. Then, click on the sketch line to recreate the path as indicated by the green arrow.

4. Click **OK** to apply the changes.

You now have a completed S-Hook with smooth curves. You can go back into the sketch and readjust any dimensions, such as the overall length from something such as 3 inches to 4 inches or the width from 2 inches to 3 inches. Experiment and see what breaks and what works. You may notice that some of the fillets that were added yield some odd results. This is due to the dimensions and how the sketch was created. Sometimes, the most straightforward way to use a simple line and fillets may not lead to the best result. Let's try another experiment where we use the Arc and Sweep tools instead.

Creating an S-Hook using the Arc and Sweep tools

We will start with a new component within the existing S-Hook file that we've worked on. Be sure that this new component is started on the top level of the assembly (the red arrow in *Figure 4.35*) and not within the other S-hook. You can do this by moving the mouse over the top of the browser project name and clicking on the dot that appears at the end. The dot shows the active component.

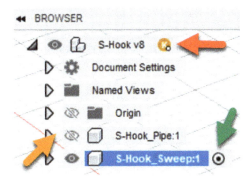

Figure 4.35 – The green arrow showing the active level dot

1. Start a new component by left-clicking on the **Component** icon ( ) and naming it S-Hook_
    Sweep. The active dot will now be moved from the top level to the new component to show
    that this is now the active level. Click on the eyeball icon and turn off the S-Hook_Pipe
    component as indicated by the orange arrow in *Figure 4.35*.

2. Start a new sketch in the **XY** plane as was done in the previous example.

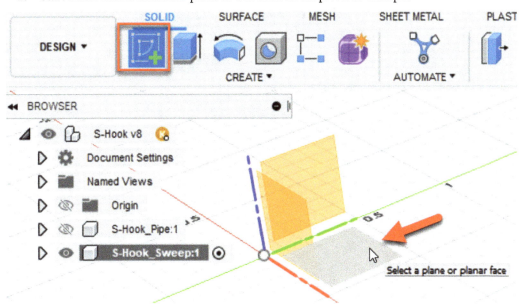

Figure 4.36 – The location of the sketch button

3. Draw a vertical line away from the origin dot to anywhere on your screen; the size doesn't matter. We will manually place this line from its midpoint to the origin point. If the line is not vertical, that is okay; just click on the **Horizontal/Vertical** constraint button (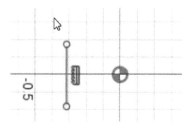) and then click on the line.

Figure 4.37 – The location of the line

4. Click on the midpoint constraint to activate it, then click on the line, and then click on the origin dot.

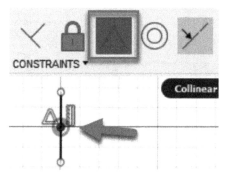

Figure 4.38 – The midpoint constraint

5. Start the Line tool and draw a horizontal line from the top endpoint to the right. Right-click on this line and then set it to **Normal/Construction**. You can alternatively click on the line and then hit the *X* keyboard shortcut.

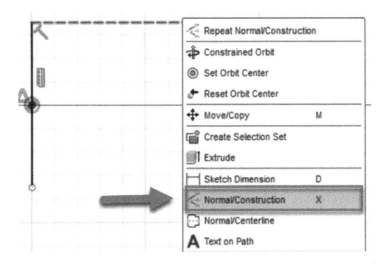

Figure 4.39 – The construction line

6. Repeat this process for the bottom endpoint of the vertical line but instead draw the line to the left. Be sure to set it to construction by selecting the line and then right-clicking and selecting **Normal/Construction** or by selecting the line and hitting the keyboard shortcut *X*. Then, either set the line to horizontal or use the perpendicular constraint, as indicated by the red arrows in *Figure 4.40*.

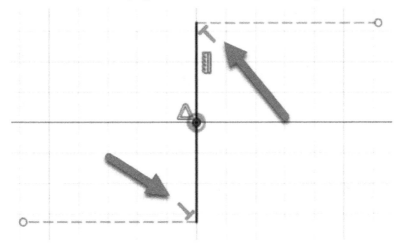

Figure 4.40 – The construction lines with a perpendicular constraint

7. Go to the **CREATE** panel dropdown and mouse over the **Arc** tool and then left-click on the **3-Point Arc** command.

Note the following about the Arc tool:

i. To place a 3-point arc, you will need to place a start, then the end location, then the top of the curve point.

ii. Click to place your start point (indicated by **1** in *Figure 4.41*), left-click to place the endpoint (**2**), and then finally place the top of the arc (**3**).

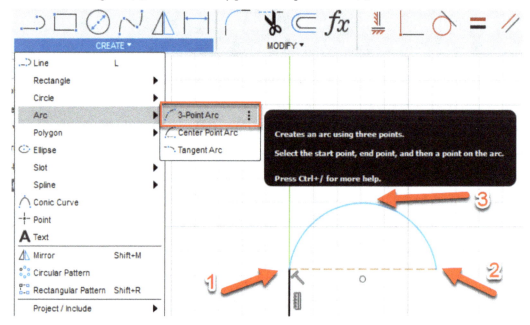

Figure 4.41 – The location of the Arc tool

8. For the bottom arc, go to the **CREATE** panel, then mouse over the Arc tool, and select **Tangent Arc** this time. Left-click once on the endpoint nearest to the line (as indicated by **1** in *Figure 4.42*) and then left-click the opposite end (**2**).

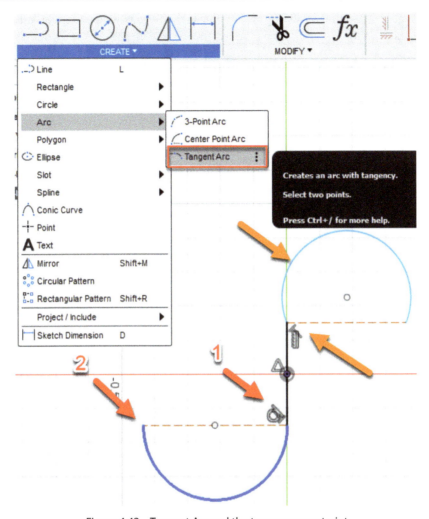

Figure 4.42 – Tangent Arc and the tangency constraint

Note the following about **Tangent Arc**:

- When placing a tangent arc, Fusion 360 will automatically place a tangency constraint next to the line selected for tangent relation

- Having tangency means that if this line gets larger, it always maintains tangency at the location indicated by **1** and the red arrow in *Figure 4.42*

- To visually see the difference, left-click and drag the top arc and notice that it gets bigger but it bubbles out and loses half its shape

- Left-click and drag the bottom arc and notice that it maintains its shape, as indicated by the orange arrows

9. Set a tangency constraint for the top arc by going to the **Tangency Constraint** button and then clicking once on the line and then on the arc.

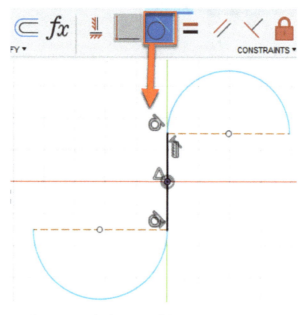

Figure 4.43 – The location of the tangency constraint

This hook looks much better than the previous hook example and has fewer points of possible errors when we try to flex the shape to try various sizes. Let's add the dimensions.

Adding dimensions

Now that we have the general shape of our new S-Hook, we can add the dimensions:

1. Use the *D* keyboard shortcut to start the **Dimension** tool and left-click on the line to place a dimension.

   There are some things to note about the **Dimension** tool:

   - It's a three-click process. Click to place the one end, then click to place the opposite corner, and click again to place the dimension location.

   - You can select the line itself, move the mouse to the left or right, and then finally left-click to place the dimension.

   - You can either accept the dimension as is and then change it afterward or change it when it is placed.

   - Set **Distance** to a value of . 5 by double-left-clicking on the text after the dimension has been placed or while you are left-clicking the final value; either way works.

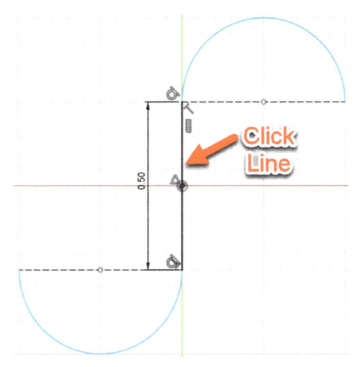

Figure 4.44 – The line dimension

2. Start the **Dimension** tool once again and left-click on the top arc to place a radial dimension:

    •   Remember, you can either set the value now by typing in the number and hitting *Enter* or wait until after you have placed all dimensions and then go back and double-click on the dimension text to start changing the dimensions.

    •   You can reference other dimensions so that when one changes, so does the other. We can match the top and bottom arcs to have the same dimensions.

    •   When you place the dimension for the bottom arc, left-click on the arc, and then place the dimension.

    •   However, instead of typing in a value, left-click on the top radial dimension and notice that when you left-click on it, it will show a value that is similar to the top arc and also add in a bit more text, which is `fx:`.

    •   If you notice that a dimension is labeled `fx:`, this means that this dimension is now referencing another dimension. If you ever want to make both arcs different values, you can just double-click on the `fx:` dimension and change it to another value.

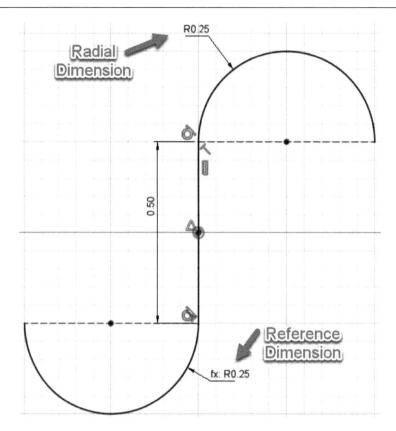

Figure 4.45 – The radial dimensions and the reference dimension

3.   Finally, left-click on the **FINISH SKETCH** button.

We will now use a sweep to create the 3D model of the S-Hook to avoid being limited by the three shapes within the Pipe tool and use any custom shape we want.

Using the Sweep tool to create a 3D model

The **Sweep** tool is used to create straight or curved 3D geometry such as a paper clip that you may use at home or at work. The way the Sweep tool works is by creating a path and then selecting another sketch that is parallel to the path. It's similar to the Pipe tool in the previous example but you can use any shape for the profile, not just the default three. The Sweep tool has a few rules that you need to know though in order to get the results that you want.

Here are the rules to note about the Sweep tool:

- The first rule is that the path must intersect with the profile. This means that a sketch plane needs to cut through the path at some point, usually at the end or the middle (see *Figure 4.46*).

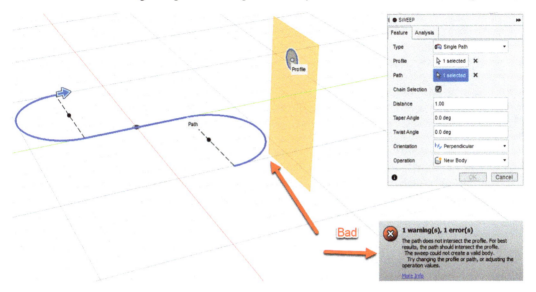

Figure 4.46 – A profile not cutting through the sweep path

- The next rule is that the size of the shape depends on the curves within the shape:

  - If your shape is too large, it will not be able to make the turn in the path, sort of like a large Mack truck not being able to make a small turn without hitting something. A small car can easily make the turn, so be sure to adjust the curve shape accordingly.

Let's get working by creating another S-Hook but using the Sweep tool instead. We first need to create the profile sketch to go along with the paths that we created previously.

Creating the profile

We already created the path for our sweep so now we just need to create the profile. Let's use the **Polygon** tool to create a multisided sweep:

1. Start the **Sketch** tool and select the **XZ** plane. See *Figure 4.47* for the plane's location.

Figure 4.47 – The XZ plane

2. In the **SKETCH** environment, click on the **CREATE** panel dropdown, mouse over the **Polygon** tool, and then left-click on **Inscribed Polygon**. Left-click to place it on the origin and move your mouse to place the point of the inscribed polygon facing up and down. You can change the number of sides if you like or leave the default of **6**.

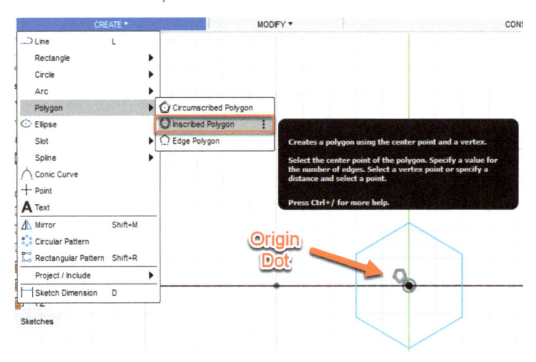

Figure 4.48 – The Polygon tool and sketch placement

3. The polygon shows up as blue, meaning that it is not fully constrained. If we were to left-click and drag the polygon, you would notice that it rotates. Let's add some constraints to lock down this rotational movement so that it stays horizontal/vertical. Left-click on the **Horizontal/Vertical** constraint button in the **CONSTRAINTS** panel and then left-click on the right edge of the polygon.

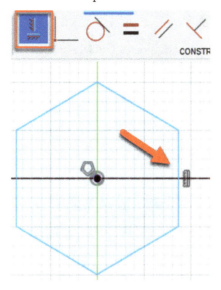

Figure 4.49 – The location of the horizontal/vertical constraint

4. Add a dimension, set it to .10, and notice that everything turns black even though we only added one constraint. That's because the polygon has its own constraints added in.

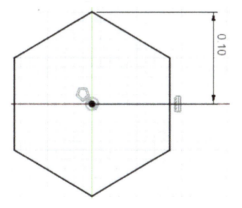

Figure 4.50 – A fully constrained polygonal sketch

5. Click on **FINISH SKETCH** to exit back into the **DESIGN** workspace.

Now that we have both the sketch profile and the sketch path set up, we can use the Sweep tool.

Using the Sweep tool

Now that both the profile and sweep sketches have been created, we can put them both together to generate the 3D model of our S-hook:

1. Left-click on the **CREATE** panel dropdown and left-click on the Sweep tool.

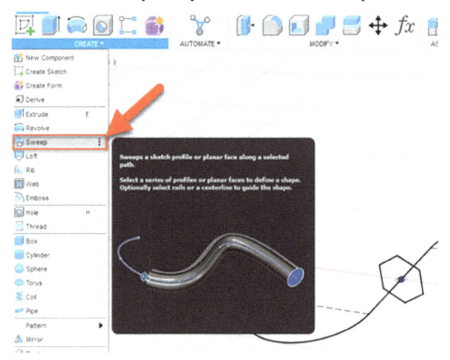

Figure 4.51 – The location of the Sweep tool

2. The **SWEEP** tool pop-up window will appear on the right side of your screen. Left-click on the area that asks for the profile and then select the polygon sketch profile that you created. Then, left-click on the **Path Select** button and select the path.

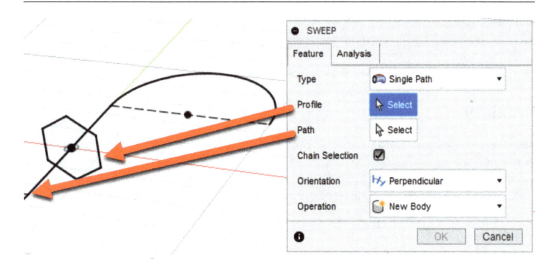

Figure 4.52 – The SWEEP pop-up window

3. You will get a few more options within this window after you have created the 3D model. Experiment more with any of these options if you'd like such as **Twist Angle**, **Distance**, **Taper Angle**, and so on. If you are happy with the results, click on **OK**.

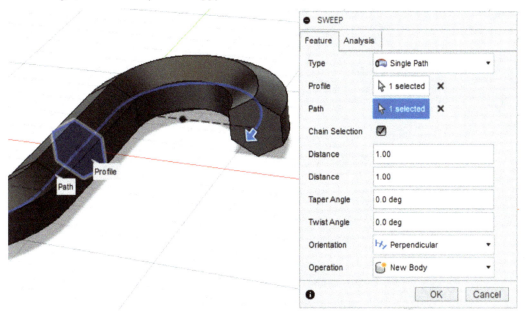

Figure 4.53 – The final sweep result and more options within the pop-up window

You have finished the S-hook and have hopefully noticed that there is more than one way to create a hook. This is why it helps to determine the design intent of the model. Is it just going to be a quick model or will you require a lot more variety in the shapes? Always spend time planning out a strategy, as this will help save you time while you are creating.

Summary

In this chapter, we learned how to create an S-Hook using two different methods. The first just used the Line tool and the Pipe tool. The second used the Arc tool along with the Sweep tool. Both methods are perfectly fine, but some have design limitations, which are necessary to know before getting started on a project. In the next chapter, we will learn how to create decorative doorknobs using the **Revolve** tool.

5
Designing Decorative Doorknobs

As you learned in the last chapter, there is more than one way to create something within Fusion 360. In this chapter, we will learn how to create multiple decorative doorknobs using the **Revolve** and **Extrude** tools. These are two of the most commonly used tools used in any 3D modeling software, and in this chapter, we will focus on various ways to use them. We will learn about three different methods: *revolve by adding*, *revolve by subtracting*, and *build upon extrusion*. These are all great ways to create 3D models, and they all have different design intents. It's up to you to determine which works best for your project, but ultimately, the process really is determined by project time and budget.

By the end of this chapter, you will have learned three different ways to build a model, which will expand your view of how to create 3D objects. This will help to show you that there is more than one way to create an object, and sometimes the most straightforward way may not be the best or the fastest.

In this chapter, we're going to cover the following main topics:

- Creating a model by adding material using the Revolve tool
- Creating a model by removing material using the Revolve tool
- Creating a model by stacking with the Extrude tool

Technical requirements

You can practice with the files provided, or feel free to create your own for a more customized experience. The sample design for this chapter can be found at https://github.com/PacktPublishing/Improving-CAD-Designs-with-Autodesk-Fusion-360/tree/main/Ch05.

Creating a model by adding material using the Revolve tool

In this section, we will learn how to use the Revolve tool by creating a sketch profile of a doorknob model and then revolving that sketch around a central axis. We will employ the method of material addition (see *Figure 5.1*). There are many real-world items that can be created this way, such as a coffee cup, a lamp, or a doorknob, which is what we will be working on today.

Before we get into any of that, though, let's first understand the design intent of what we are creating. Will this piece be intended for manufacturing or 3D printing, or will it be used for 3D rendering? Let's assume that this piece will be used for 3D rendering. This way, we can concentrate on the usage of the Revolve tool. We will get into more details about manufacturing and 3D printing in later chapters.

Creating a sketch profile

Now that we understand what we are creating this piece for, the next step is to create the sketch profile of the 3D doorknob model. To do this, we will first take a look at the finished model to see what we will be creating so that it will help you envision what will be needed to create the side profile sketch.

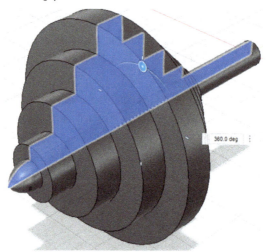

Figure 5.1 – One of the finished models

One way to create this doorknob is to create a side profile of it and then use revolve around the center axis:

1. Start a new Fusion 360 drawing:

    a. Click the **Save** button, create a new folder within the Packt Pub project, and name it Ch5_Doorknobs.

    b. Save the file and name it Doorknobs.

This is basically the same project setup as in the previous chapter except with new names. The project names don't really matter while you are working through this book, but they are helpful for you to locate what project you are working on. Feel free to name these projects however you wish; just remember where you placed them, and name them something similar to what the project is.

2. Let's stick with rule #1 and create a new component and then name it Doorknob_Revolve_ Addition.

3. Click on **Create Sketch** and select the **YZ** plane. The reason why we are choosing the **YZ** plane is to have the doorknob relative to how it will look in the real world, with the knob facing toward the screen. You could also similarly choose the **XY** axis since it will result in the same shape, facing the same direction.

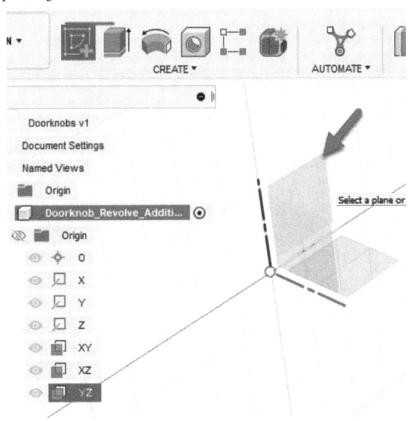

Figure 5.2 – The YZ plane

4. Create a shape similar to the image shown in *Figure 5.3* with the origin toward the back where the connection hole to the door would be. Don't worry about the length or dimensions just yet; just concentrate on the general shape. This is something that you typically do when working on a Fusion 360 project: work on the general shape first, then add in the constraints and dimensions.

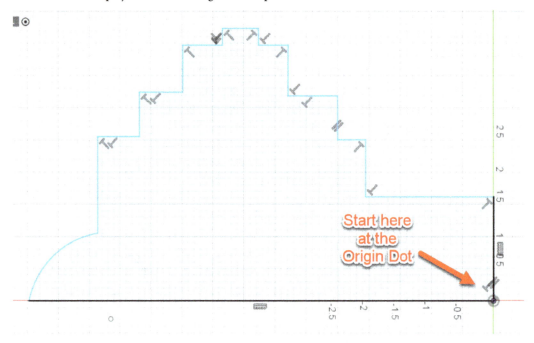

Figure 5.3 – The doorknob Revolve sketch

> **Important note**
>
> The constraints shown in *Figure 5.3* were added automatically by Fusion 360 while the sketch was being created. Your constraints may differ depending on how it was created by you. You can always add constraints by selecting them in the **CONSTRAINTS** panel and delete them by left-clicking on the *Constraint* icon in the sketch and hitting *delete* on your keyboard.

We will now add in some more constraints, such as the colinear constraint, which will hold lines on the same axis.

5. Left-click on the *colinear* constraint in the **CONSTRAINTS** panel, then click on a horizontal line such as the bottom-right line (the red arrow marked as **1st** in *Figure 5.4*), and then left-click on the separate horizontal line directly across from it (the red arrow marked as **2nd** in *Figure 5.4*). Now, if one line moves on the *y* axis, the other line will move with it.

6. Repeat for the remaining lines according to *Figure 5.4*.

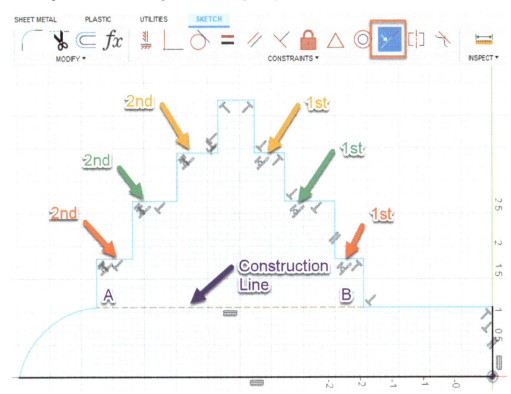

Figure 5.4 – The colinear constraint locations and construction line

Now, we will maintain a horizontal relationship between the bottom line on the right and the arc dot on the far left (see the construction line in *Figure 5.4*).

7. By adding a construction line instead of an actual line, we can show a relationship without creating a profile, which can get in the way sometimes when creating a solid model.

8. Left-click on the Line tool or hit *L* for the keyboard shortcut, then press *X* to add a construction line from the arc endpoint dot (point **A**) to the endpoint dot opposite it (point **B**) (see the purple arrow in *Figure 5.4*).

9. To view a list of the most current shortcuts, click on this link: https://www.autodesk.com/shortcuts/fusion-360.

To turn a line into a construction line or vice versa, you can either select it, go to the **SKETCH PALETTE** options on the right side of your screen, and hit the **Construction** icon, or you can use the keyboard shortcut *X*. You can also right-click on the line and choose **Normal/Construction** from the pop-up window.

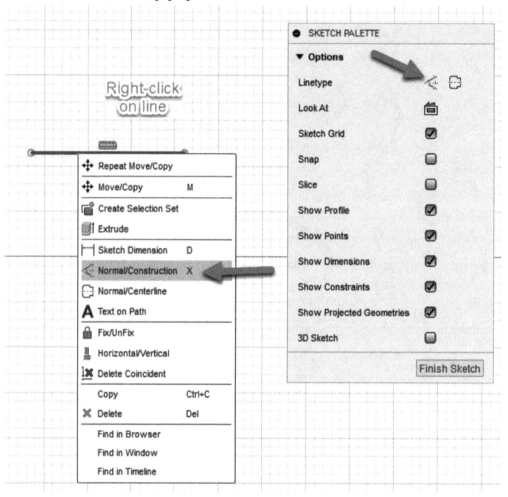

Figure 5.5 – Right-click shortcut menu construction line tool's location and SKETCH PALETTE

10. Left-click on the *equal* constraint and select the lines shown in *Figure 5.6.*

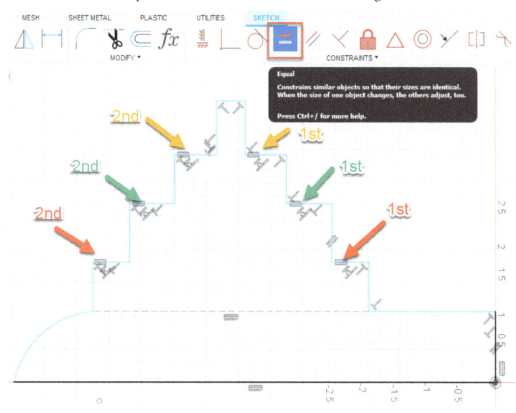

Figure 5.6 – The equal constraint locations

The reason we are using this constraint is that it will set the line to have the same length on both sides, and when it is time to dimension these lines, we will only need to add dimensions to one side rather than both sides.

This should be enough to start adding in dimensions. Typically, when first adding dimensions to a sketch, I add in the overall length and the overall height, which then shrinks or expands the sketch so that the rest of the sketch objects are closer to the correct size. This also locks the down the sketch a little so that lines possibly won't move around as much.

After adding in my first overall dimension, I noticed that I created my sketch too large. Typically, you want to create 3D models whose size is as close to the real-world size, even if it may not be used for production. So, I did a quick Google search and found that a typical real-world doorknob is around 2 inches wide and 3 inches tall. I will need to use the **Sketch Scale** tool to bring my sketch closer to actual real-world dimensions.

> **Important note**
>
> It is good practice not to scale the 3D model down if you make a mistake in the sketch. The sketch is the object that drives the parameters of the 3D model, and if you scale down the 3D model, you may get unintended results. Try to match the sketch to the 3D model. If, though, you received a 3D model from an outside source with no sketch, then it is perfectly fine to scale that model.

Adjusting the sketch size using the Sketch Scale tool

Since we started off without adding any dimensions, which is typical when creating sketches, if you created your sketch too small or too large, as I did in *Figure 5.7*, and want to scale it all down, you can use the **Sketch Scale** tool located within the **MODIFY** panel.

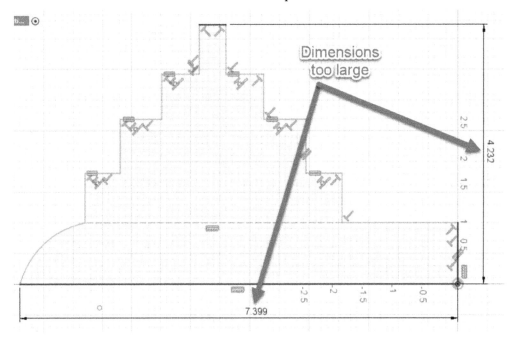

Figure 5.7 – Dimensions are too large for a doorknob

You will need to scale at a specific point (usually, it will be the origin point), and you will need to scale by a percentage. This means if you want to double the size, you will need to scale by 2, and if you want to scale by half, you will need to scale by .5.

Let's try to fix the overall size of the sketch by using the **Sketch Scale** tool:

1. In the sketch environment, go to the **MODIFY** panel and then click on the **Sketch Scale** tool.

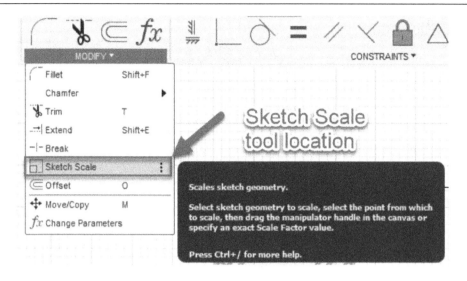

Figure 5.8 – The Sketch Scale tool's location

2. A flyout window will open asking you to select entities. You can either use your keyboard and hit *Ctrl + A* to select all or you can use your selection window and left-click in the top-left corner, then drag to the bottom-right corner, over your sketch entities.

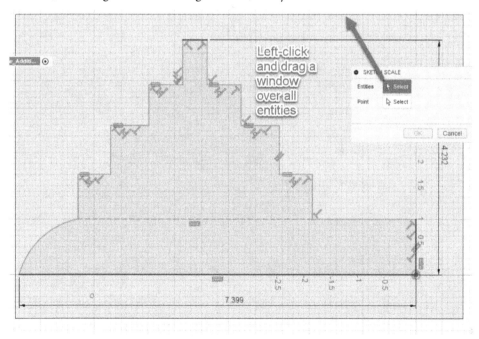

Figure 5.9 – The selection window over all entities

3.   The Entities selection will show 41 selected

   a.   We now need to select a point on the sketch, as this will be the location the scale action will grow or shrink from.

   b.   Left-click on the **Point Select** box. Some Fusion 360 tools will automatically select the next step for you, while others will not. In this case, it does not since you could possibly keep selecting objects, and Fusion 360 will not know when you are done selecting. You will need to manually move to the next part, as it will not move down to the next selection for you.

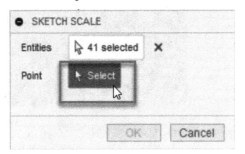

Figure 5.10 – The Point Select box

4.   Once you have selected the **Point Select** box, left-click on the origin point, which is the bottom-right corner point of your sketch.

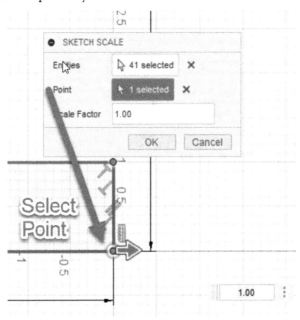

Figure 5.11 – Location of the origin point for point selection

5. Notice that a new option opens asking for the scale factor. This works by percentage, which is why it starts off at 1, meaning this is at full size. We need to make it smaller:

    a. Since we want to make the overall length 3 in, we can take the current number (my overall length is 7.399, shown in *Figure 5.7*, but yours may be different) and then divide it by 3 in.

    b. In the box that shows `1.00`, type in the `3/7.399` formula (or whatever your number may be). Notice that the sketch will be scaled down immediately.

    c. Click **OK** to finish the command.

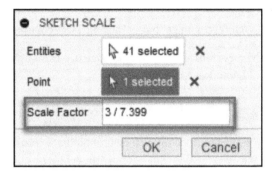

Figure 5.12 – Location of the Scale Factor input

6. Now add in the dimensions, as shown in *Figure 5.13*, to finish off the sketch.

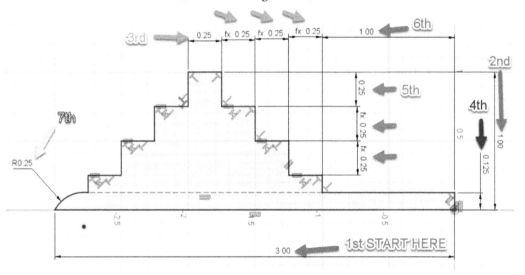

Figure 5.13 – The dimensions of the sketch

I used the order in which I normally place dimensions as a visual for you, but you don't have to follow this order. I typically like to do the overall dimensions first, then add in the smaller ones after. Also, notice for the **3rd** and **5th** dimensions that the letters fx appear next to most of them. This is called a "driven dimension" because one dimension is controlling another. You can do this too by first placing a dimension (see the orange arrow in *Figure 5.13*), and then, when placing the subsequent dimensions, instead of typing in a number, left-click on the first dimension that was placed, then left-click away to place the dimension and a small fx will appear next to it. Then, when placing the subsequent dimensions, instead of typing in a number, left-click on the original dimension.

7.   Click on the **Finish Sketch** button to get out of the **SKETCH** environment.

Now that you have added the dimensions and finished the sketch, we can finalize the project by using the Revolve tool to create the 3D model.

Finalizing the model with the Revolve tool

After clicking the **FINISH SKETCH** button, you will be brought back into the **DESIGN** workspace. We can now use the sketch that we created to build our 3D model of the doorknob:

1.   Click on the **CREATE** panel drop-down arrow and then left-click **Revolve**. The Revolve tool is also located in the shortcut buttons at the top of the screen by default.

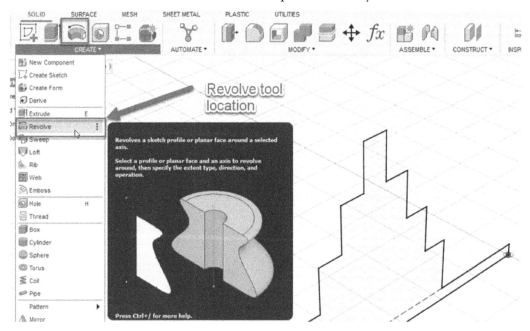

Figure 5.14 – Location of the Revolve tool

2. The **REVOLVE** flyout will open to the right side of your screen and usually will select the last profile that was created; if not, select the profile.

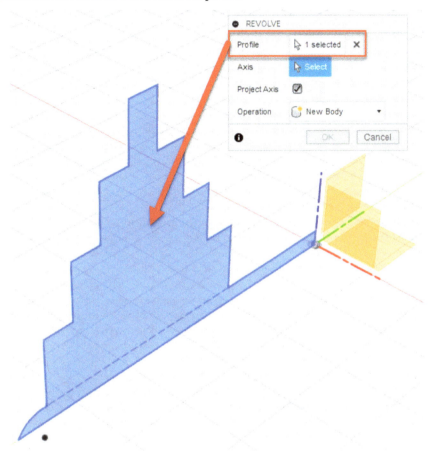

Figure 5.15 – The selected profile

Important note

Be sure that there are no gaps in your profile; otherwise, you won't be able to generate a 3D object. You can tell that you have a gap if your profile is not selectable and it does not show up as light blue, as shown in the following screenshot:

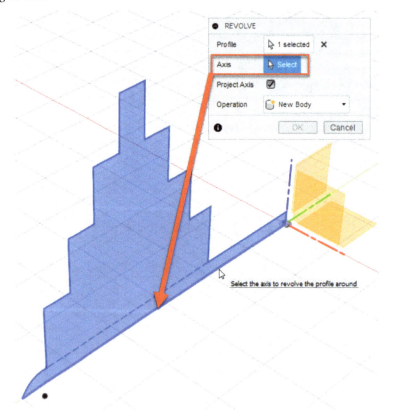

3. Select the **Axis Select** button and click on the long line at the bottom of the sketch, as shown in *Figure 5.16*.

Figure 5.16 – The Axis Select button

4. A preview will be generated showing what the 3D model will look like with the current settings, and a new pop-up window will open showing more options. Most of the time, these settings are the ones that are most used, and you can click the **OK** button to finalize the project, but feel free to play with some of the settings, such as the angle, by setting it to 30 degrees, and see what it looks like. You can also change the axis to any horizontal or vertical line to generate a completely different shape!

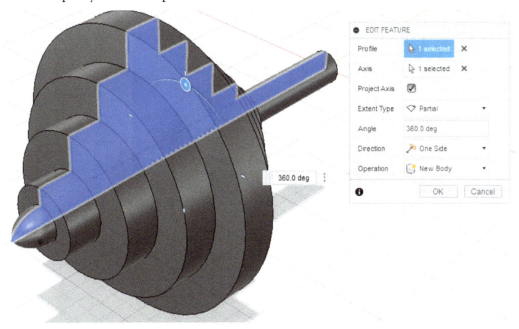

Figure 5.17 – The final doorknob

The doorknob is finished, but let's make some further changes, such as removing the pointy decorative top, making it much smoother, and adding the door hole cover since this is just a model for rendering and not for manufacturing. If this model were for manufacturing, you would need a separate center piece, a separate doorknob, and a separate door hole cover, which would all be put together as a multi-part component.

Modifying the sketch

Often, after you have created a 3D model, you may notice some things that you might not like about it and will want to change. This is where the true power of having a parametric sketch control your 3D model comes into play. I have many years of experience using AutoCAD and wish I had the ability to quickly change a 3D model that I thought would look perfect but, in the end, did not turn out the way I thought it would. With Fusion 360, you now have this amazing ability:

1. To edit the sketch and make further adjustments, right-click or double left-click on the sketch icon in your timeline and then go to **Edit Sketch**. You can also similarly go to **BROWSER** on the left side of your screen and double left-click on the sketch or right-click on **Edit Sketch**.

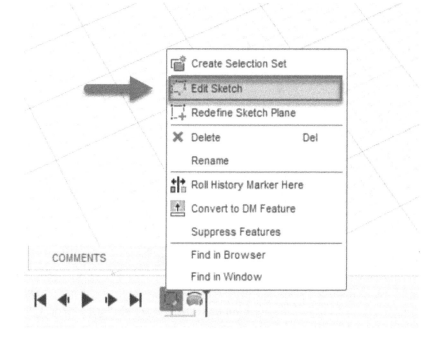

Figure 5.18 – The Edit Sketch icon location

With the sketch open, we want to make the front decorative pieces less sharp (and a bit more smooth). To do this, we must free it from the construction constraint that is holding it in place at the bottom endpoint.

2. Left-click on the construction line and then hit the *delete* key on your keyboard. We could also, alternatively, select the coincident constraint endpoint on the right side of the construction line, but sometimes it's easier to just completely remove the line and then add it back.

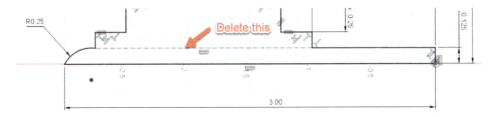

Figure 5.19 – The construction line to remove

Notice that the front circle is now light blue in color, allowing us to make the arc larger.

3. Left-click on the white dot of the arc and drag it around to see the size changing.

4. Remove the radius dimension as well. We will be using a construction line to control it.

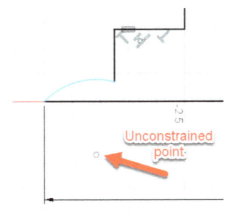

Figure 5.20 – The radius center point

5. Start the Line tool and add a construction line from the center point to the endpoint just above. Now add in the long horizontal construction line, but instead of the endpoint on the right, connect it to the midpoint of the vertical line. Notice that all lines are now black, which means that the sketch is fully constrained once again.

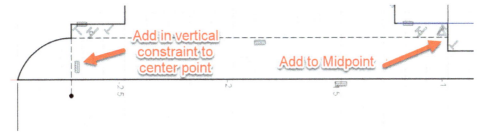

Figure 5.21 – The added construction lines

6. To add the door cover, we will need to add a piece for the back:

    a. First, we will need to remove the rear line (See *Figure 5.22*) but keep the dimensions. Notice that the internal blue color of the sketch disappears, meaning that the sketch is not selectable to generate a 3D model. If you were to click on **FINISH SKETCH** now, you would generate some errors.

    b. Also, notice that some lines turn blue while others remain black. This means some lines are movable, the light blue ones, while others are still being constrained due to the dimensions placed on them.

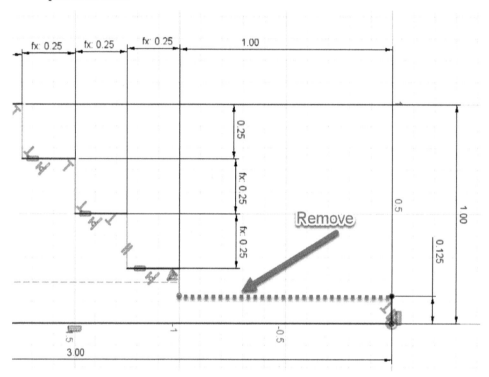

Figure 5.22 – Showing the rear line removed and the sketch opened

7. A typical rear doorhandle cover is about 2.5 in in diameter, but since we are generating a 3D model by revolving around the model's origin, we only need to make it 1.25 in since 1.25 in x 2 in equals 2.5 in:

    a. Instead of typing in 1 . 25, we can use a formula within the dimension and let Fusion 360 do the math for us.

    b. Double left-click on the vertical .125 dimension, change it to 2 . 5/2, and hit *enter*.

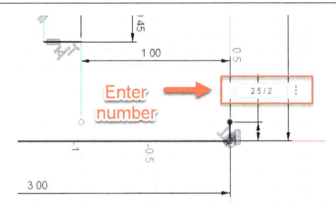

Figure 5.23 – Formula to enter for the dimension line

Notice that the line changes length, and the dot on the left also moves up with it. This is because the dimension that we placed earlier was connected to the dot rather than the line.

8. To fix this, we need to remove the 1.00 in dimension so that it can be dragged back down to its proper location.

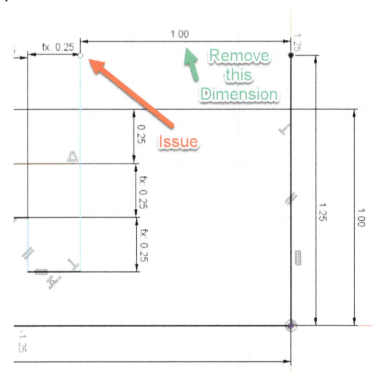

Figure 5.24 – The new dimension and an issue with a dot

9. With the dimension removed, you can now left-click and drag the white dot down to roughly its original location. We will place it in the correct location once we have added the new lines for the doorknob cover plate.

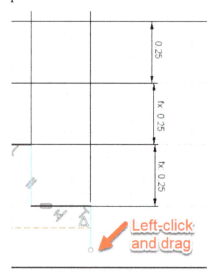

Figure 5.25 – The new location of the white dot

10. Left-click on the Line tool and draw in the lines to close the opening and start creating the shape for the doorknob cover, as shown in *Figure 5.26*.

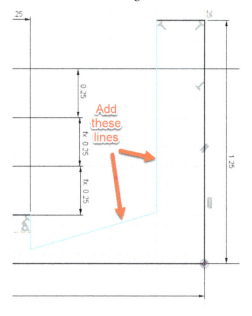

Figure 5.26 – The lines for the door cover

We need just the rough shape outline and then we can add in the constraints afterward. Notice that since we have added these lines in, the internal light blue color returns, signifying that we can use this shape to generate our 3D model.

11. Add in the constraints as shown in the image by left-clicking on the horizontal/vertical constraint and then left-clicking each line.

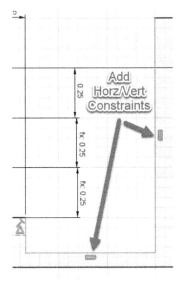

Figure 5.27 – The newly added constraints

12. Hit the *D* keyboard shortcut and add in the dimensions as shown to complete the sketch.

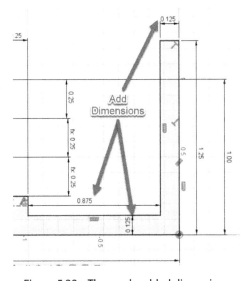

Figure 5.28 – The newly added dimensions

13. Click **FINISH SKETCH** to go back into the **DESIGN** workspace to see the finished results.

Figure 5.29 – The finished doorknob with door cover backplate

We have finished the first model and have updated it with a doorknob cover on the back by simply going back into the original sketch and updating it with a few changes. Now we will create another doorknob by first creating a simple cylinder and then removing material similar to how a woodworker uses a lathe.

Creating a model by removing material using the Revolve tool

In the previous example, we created a sketch and then used the Revolve tool to generate a 3D model. This time, we will create a 3D model and then remove material from the model to create the same basic shape:

1. Continue within the current doorknob drawing by creating a new component:

    a. You can do this by right-clicking on the **BROWSER** top component and then going to **New Component**.

    b. Notice that the active component is the top-level component named `Doorknobs v9` (see *Figure 5.30*), not `Doorknob_Revolve_Addition`.

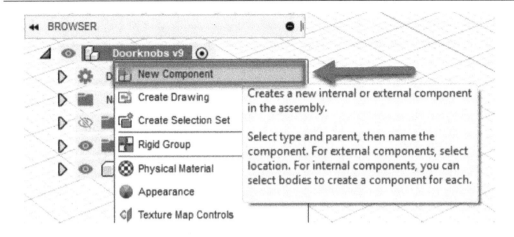

Figure 5.30 – The New Component location

2.   In the pop-up component flyout, name the new component Doorknob_Revolve_Subtract. You will see the previous model turn transparent, which is normal when you create a new component. Turn off the old model by clicking on the eyeball icon.

Figure 5.31 – The eyeball icon

Important note

If you click on the eyeball icon and your previous model is still on the screen, it may be due to the `Bodies` folder being outside of the component. You can fix this by going down to the timeline and removing the last Revolve, which will also remove the `Bodies` folder. Now make sure the `Doorknob_Revolve_Addition` component is active and then recreate the Revolve, which will place the `Bodies` folder in your current component. Components act like large storage bins, as in, they can hold a variety of objects within them, and if the component that you are working on is not active, it will place a folder outside of the component. It's not going to hurt the model if it's placed outside of the component, but it can make for a messy timeline.

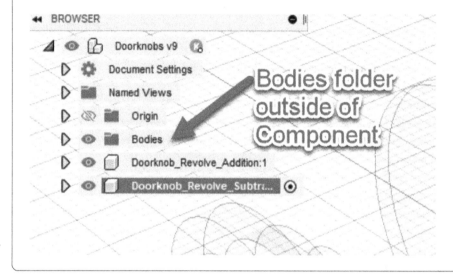

3. With the new component active, create a cylinder model by going to **CREATE** and then the **Cylinder** tool.

4. Choose the front **XZ** plane as the sketch plane.

5. Give the new cylinder the same dimensions as the previous doorknob, which is a diameter of 2.5 and a height of 3 in.

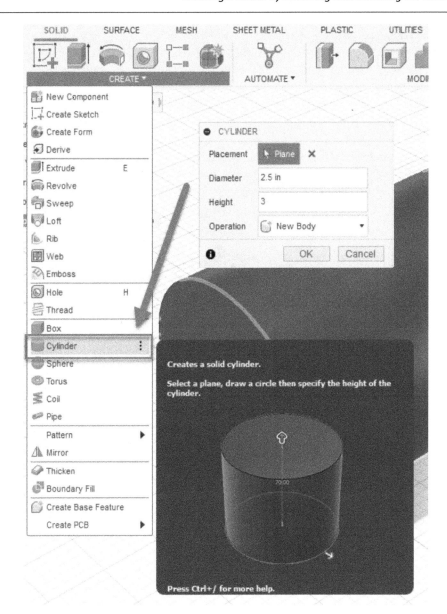

Figure 5.32 – The Cylinder tool location

The first step has been completed by generating a 3D cylinder. You could also similarly create a sketch with a circle and then extrude it, but the approach we followed creates the sketch/model in one easy step. Next, we will create the sketch that will be used to cut our model.

Creating the reverse sketch

With our model created, we can now make cuts in it. We can either do this with one sketch or using multiple sketches depending on the complexity of the model. Since this model is simple, we can use just one sketch:

1. Start the Sketch tool and select the **YZ** plane.

    If you have trouble selecting this plane, you can left-click and hold over the plane that you wish to pick, and a pop-up selection filter menu will appear, allowing you to left-click on the **YZ** name and select this plane.

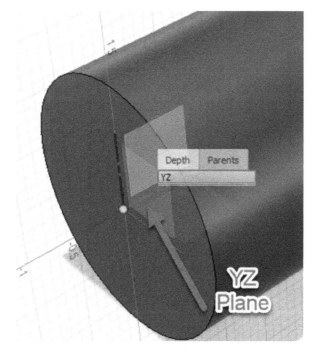

Figure 5.33 – The pop-up selection filter menu

2. Draw in roughly the same sketch shape as we did before for the previous example (I added in a few more levels to differentiate the two models).

3. Then, add in the same constraints as in the previous sketch, such as the *equal* constraint for the line lengths and *colinear* to keep lines on the same level.

4. The remaining sketch lines can either be horizontal/vertical or perpendicular (see *Figure 5.34*).

5. Notice that we are drawing lines outside of the 3D model as well. This is to make sure that when we cut away the extra material, it will all be removed without leaving any small pieces.

Figure 5.34 – The sketch shape used for removing material

6. Now add in the dimensions as shown in *Figure 5.35*.

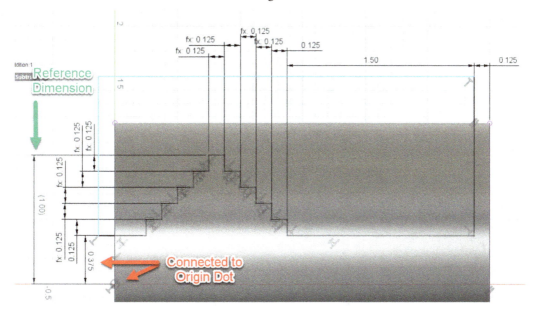

Figure 5.35 – The dimensions shown

Notice that I connected my first dimension (see the red arrow in *Figure 5.35*) to the origin dot. This will help to lock down any movement when there are any size changes to the model. Also, notice that I wanted to make sure that I didn't exceed the overall 2in diameter of a typical doorknob, so I created a driven dimension (see the green arrow in *Figure 5.35*) or a dimension that is just there for reference. The tell-tale sign that you have created a driven dimension is the parentheses around the numbers. Driven dimensions are only for reference, and if you were to try and change the size, it would not affect the model size.

7.   Click on **FINISH SKETCH** and notice that when you are brought back into the **DESIGN** workspace, you have two additional purple lines shown in your sketch.

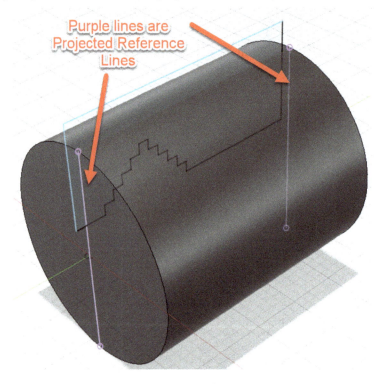

Figure 5.36 – Projected reference lines

These are projected reference lines and appear when the edge of the existing geometry (the cylinder) is selected or referenced by a dimension and pulled into the current sketch. We will be going over these a little bit more as we go along, but these are very helpful when working with existing geometry.

With the sketch now generated, we can use the Revolve tool to cut out the geometry that we don't want and keep the remaining part.

Revolve using the cut profile

The **Cut** operation is located as an option within the other **CREATE** operation tools, such as *Extrude* and *Sweep*. If you are used to working in a program such as *SolidWorks*, you would have to select a separate tool to cut the model. In Fusion 360, it is located within the options, making it easier to switch between commands:

1. Start the Revolve tool located within the **CREATE** panel, and the sketch that you created will automatically be selected. If not, click on the **Profile Select** box and left-click on the profile. Remember, if there are any gaps in your sketch, Fusion will not let you generate a 3D model from it.

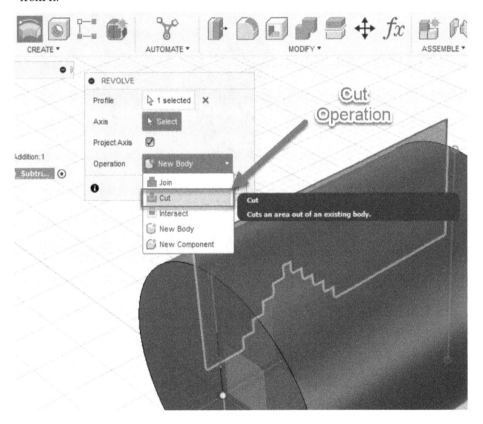

Figure 5.37 – The REVOLVE tool location

2. Select the **Cut** operation and choose the green *Y* axis (see the green arrow in *Figure 5.38*).

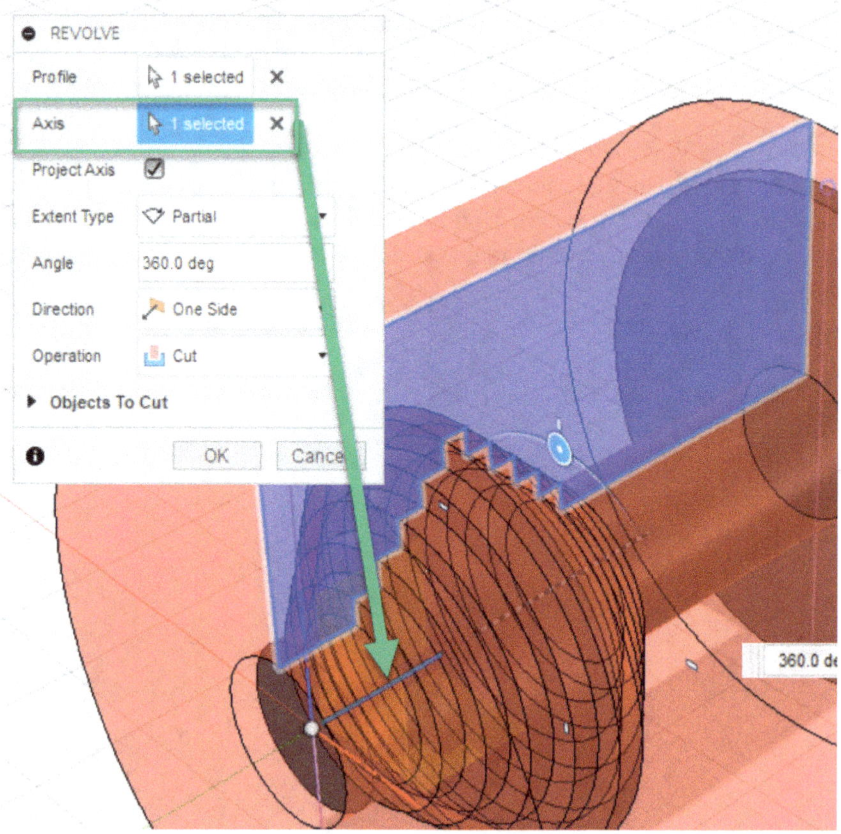

Figure 5.38 – The green axis location

3. Leave the remaining options as is and click on **OK** to complete the command.

Now that we have completed the cut operation by removing some material, we can add some fillets to smooth out sharp edges.

Adding fillets

The final part now is to add a fillet to the front of the doorknob. This will work more easily now than before since we have existing geometry to work with rather than a sketch:

1. Within the **DESIGN** workspace, select the **Fillet** tool located within the **MODIFY** panel.

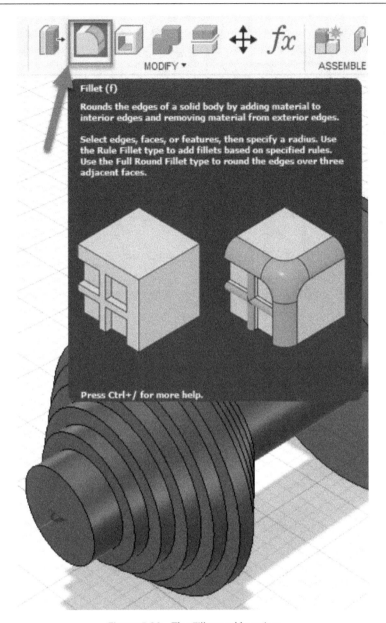

Figure 5.39 – The Fillet tool location

The **FILLET** option pop-up window will appear on the right side of the screen.

2. Select the front edge and set the distance to be .25 in.

    If you try to use more than .25 in, you will receive an error since that is the amount of space between the front line and the next line, and there is not enough material to add more.

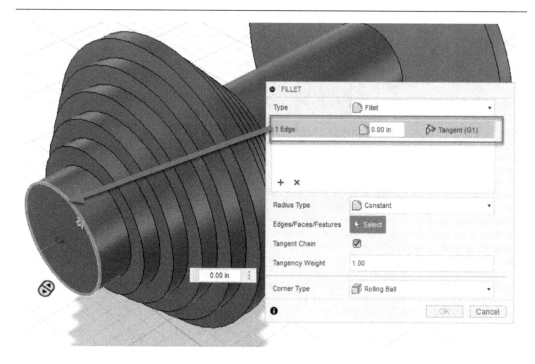

Figure 5.40 – The Fillet tool edge selection

3. Click **OK** to finalize the Fillet tool, and we have completed the *revolve with subtraction* project.

Figure 5.41 – The final revolve with subtraction doorknob

There is one problem, though, with this model. If I were to flex this model, for example, making the cylinder slightly larger, it would create external material instead of wiping out all the exterior material. This is because of the two blue lines that were not constrained earlier (see *Figure 5.42*).

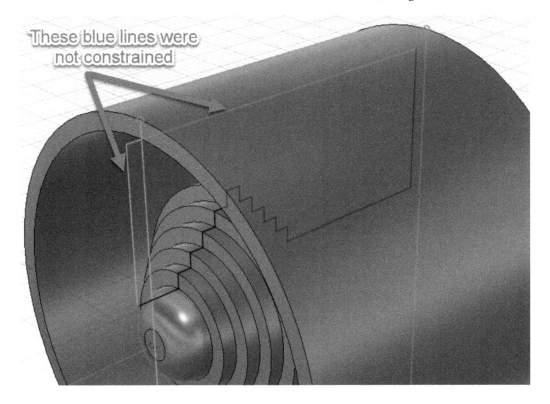

Figure 5.42 – The problem with the finished product

We were able to generate the model without any issues, but because not all the lines were black, when the model was made slightly larger, it kept some external material. We can fix this by going back into the sketch and using the Project and Trim tools.

Modifying the sketch

Just like before, we are going to go back into the sketch that we just created and make a few changes to remove the extra geometry:

1. Double-click on the *Sketch* icon in the timeline, or right-click and choose **Edit Sketch**.

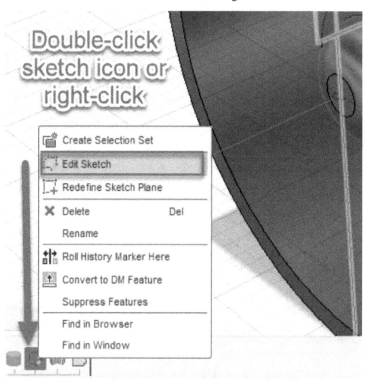

Figure 5.43 – The Edit Sketch location

2. Select the **Trim** tool within the **MODIFY** panel and trim out the three lines shown in *Figure 5.44*. Then hit *enter* to finish using the tool.

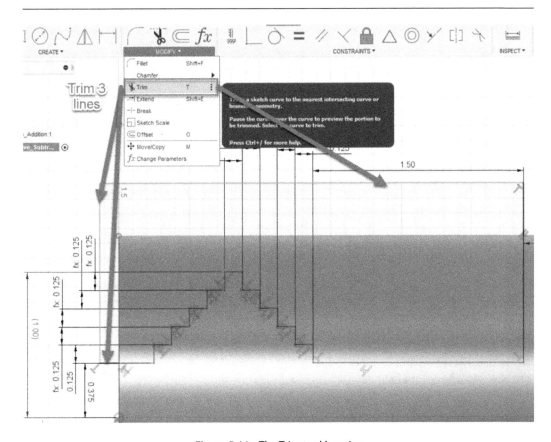

Figure 5.44 – The Trim tool location

3. Notice that once you have trimmed the horizontal line closest to the origin dot, a black dot is connected to the projected (purple) line. That line is now constrained with a coincident constraint, which means that it will not move from that location and will grow as the 3D model grows.

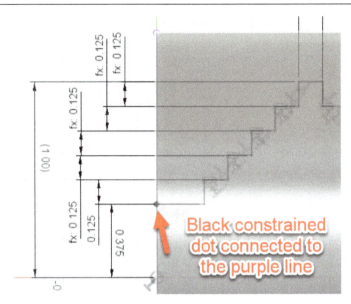

Figure 5.45 – The constrained dot connected to the purple projected line

If you were to trim the last line on the far right (see the red arrow in *Figure 5.46*), you would notice that it does not connect to the cylinder body as the other line did. This is due to the curved surface of the cylinder body and how cylinders are created in 3D.

If you ever need to draw on or use a curved edge, you will need to project that curved edge onto the current sketch.

4.   Click on the **CREATE** panel dropdown and go to **Project/Include** and then left-click on the **Intersect** tool.

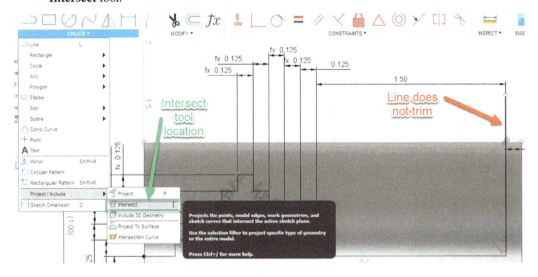

Figure 5.46 – The Intersect tool's location

5. Mouse over the cylinder body and left-click on it. Click **OK** to close the **Intersect** tool, and you will see a purple line will be generated for the side of the cylinder.

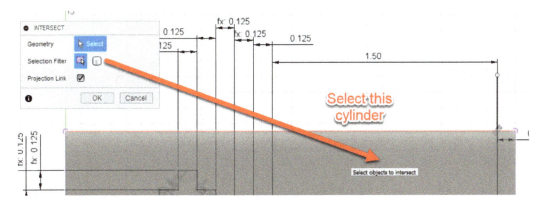

Figure 5.47 – The Intersect pick location

6. Now use **Trim** to select the last line, and it will be cut back to the projected edge.

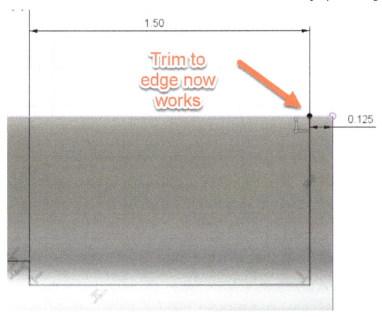

Figure 5.48 – The trimmed back line

7. Click on **FINISH SKETCH** to complete the model.

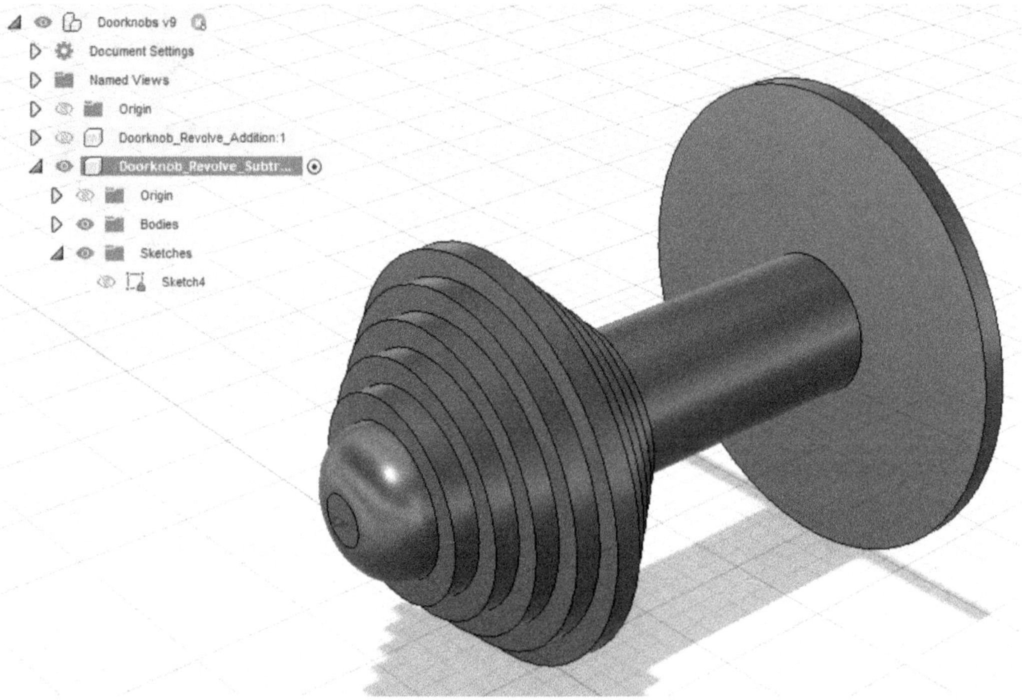

Figure 5.49 – The finished doorknob model

This is a now a better constrained model and sketch. If you were to go back into the original cylinder creation and change the size to something larger (aka flexing the model), the model would change shape without any issues.

We have created two models using the Revolve tool by (1) adding and (2) removing/subtracting material. For our last model, we will use a method of stacking cylinders on top of each other to build the 3D model.

Creating a model by stacking with the Extrude tool

In this section, we will learn how to create the same doorknob model through the method of adding material by stacking extruded shapes on top of each other. This method is another way of creating 3D models with simple shapes. Extruding a shape is a very basic form of creation in all 3D programs but is also the most used as it is very straightforward and simple. You draw a simple 2D shape, then lift that shape in a direction to form a solid body. We will be doing this multiple times to form the doorknob model in this exercise.

Creating the doorknob component

Just like before, we will continue in the same Fusion 360 project file and create a new doorknob component. Be sure that you make the top-level project active before you create the new component, or you will be creating a subassembly within the last created part:

1. Make sure that the top level is active (see the red arrow in *Figure 5.50*), then right-click on the top level of the assembly and select **New Component**. Name this new component Doorknob_Extrude.

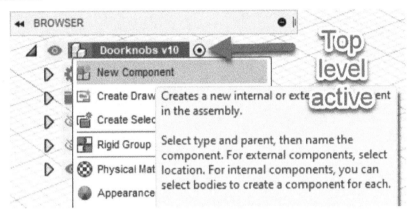

Figure 5.50 – The active top-level assembly

2. Left-click on the eyeball icon of the previous model, Doorknob_Revolve_Subtract, within the browser to hide the previous model.

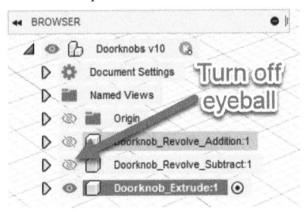

Figure 5.51 – The eyeball icon turned off

3. Start a new sketch and left-click on the **XY** plane.

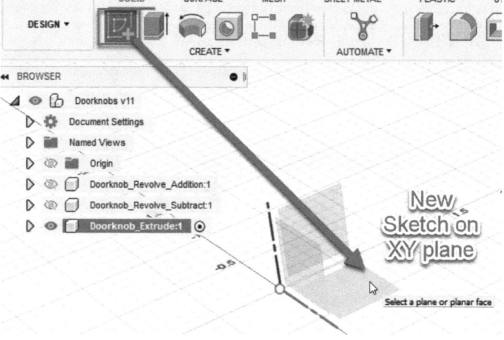

Figure 5.52 – The Sketch tool location and XY plane

We will be starting from the bottom of the doorknob, which is the door hole cover plate, and working up toward the top.

4. Left-click on the Circle tool located in the **CREATE** panel and then left-click on the origin dot to lock it in place. This time, we are making the full diameter of the circle rather than the radius. We are not going to use the Revolve tool but instead will be using the Extrude tool to generate the 3D shape.

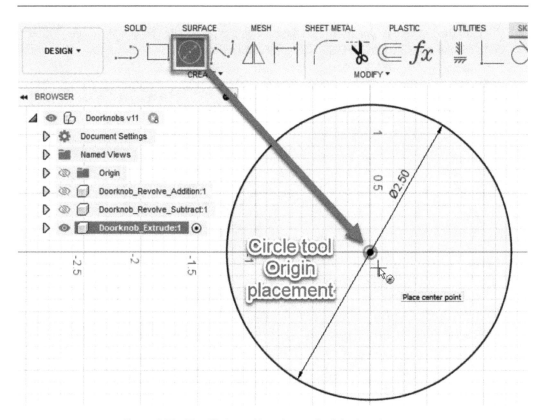

Figure 5.53 – The Circle tool location and origin dot placement

5. Click on **FINISH SKETCH** and go back into the **DESIGN** workspace.

    We will now extract that 2D shape on the Z axis to create the 3D model.

6. Click on the Extrude tool, and in the **Distance** box, type `.125`. You can leave every other option as the default and click **OK** to close the command.

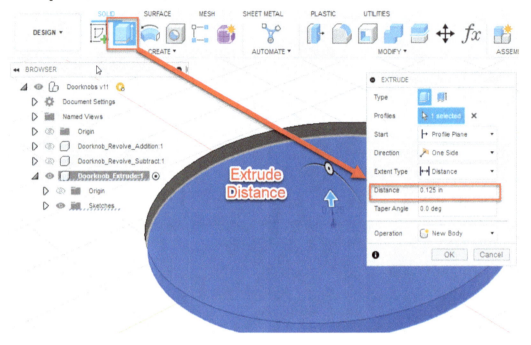

Figure 5.54 – Location of the Extrude tool and Distance

7. Left-click on the Sketch tool and then left-click on the top of the extruded part that was just created.

You can sketch on top of any flat surface without the need for a plane, and we are going to place the doorknob extension on this plane.

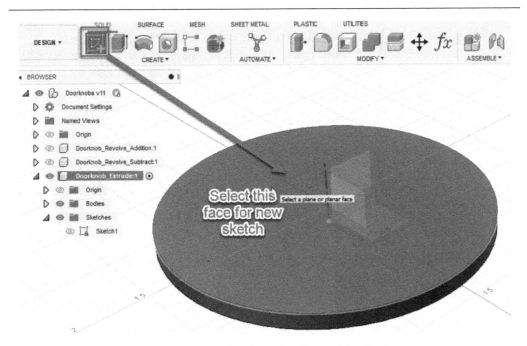

Figure 5.55 – The Create Sketch tool location and sketch placement

8. Left-click on the Circle tool within the **CREATE** panel and then place it on the origin dot. You can either type in .5 for the diameter or left-click to place the circle and then use the *D* keyboard shortcut to place the diameter.

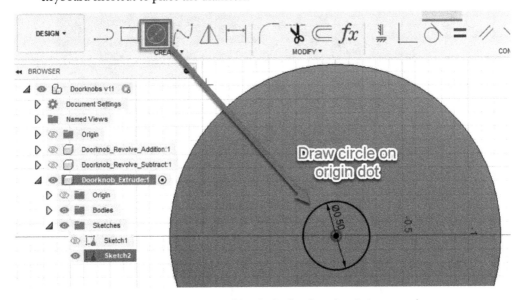

Figure 5.56 – Location of the Circle sketch tool and placement

9. Click on **Finish Sketch**, and you will be brought back out to the **DESIGN** workspace.

10. Click on the *Extrude* icon and left-click on the sketch that you just created:

    a. Type 1.5 in the **Distance** box.

    b. Be sure **Operation** is set to **Join**; otherwise, you may be creating a separate body, which is not what we want.

    c. Click on **OK** to close the command.

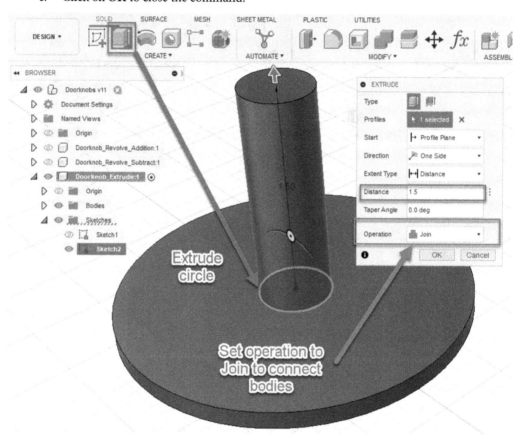

Figure 5.57 – The location of the Extrude tool and Distance

We will be repeating the process for the rest of the circles: click on a planar face of the 3D model, add a sketch, then extrude it. To save time and book space, I added a reference image for you to finish off the extruded doorknob.

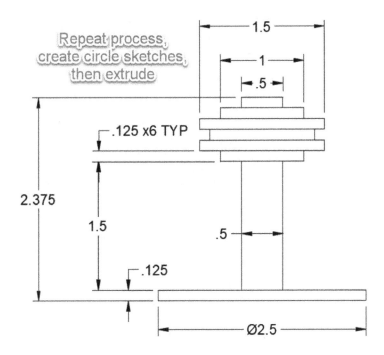

Figure 5.58 – The dimensions for the remaining circles

The overall length doesn't hit the average door length of 3", so to finish off the doorknob, we will use the Mirror tool. The Mirror tool allows us to flip and copy existing features such as bodies, faces, features, and components. It is very handy when you don't want to recreate an object and can simply mirror along a plane or a face.

Completing the doorknob with the Mirror tool

To make things go a little quicker, we will use the Mirror tool, but instead of mirroring the entire body, we will use the Feature object type to add the remaining circle extrusions. This allows us to select a tool used to build a 3D model, such as *Mirror, Extrude, Revolve*, and so on, within the timeline, and mirror that feature to create flipped and copied geometry:

1. First, let's create a midpoint plane between the last circle extrusion that we created.

   a. In the **DESIGN** workspace, go to the **CONSTRUCT** dropdown, and then left-click on the **Midplane** tool.

b.   The **Midplane** tool allows us to create a plane between two faces.

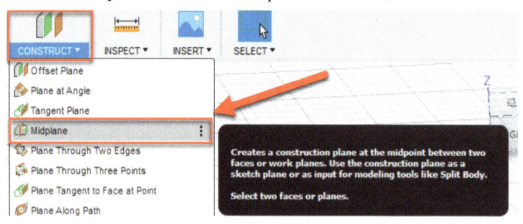

Figure 5.59 – The Midplane construction tool's location

2.   Left-click on the two faces indicated by red arrows in *Figure 5.60*. The reason why we are creating a midplane rather than a mirror on the next face is that when the mirror is created and flipped, there will be a gap in between, which is not correct.

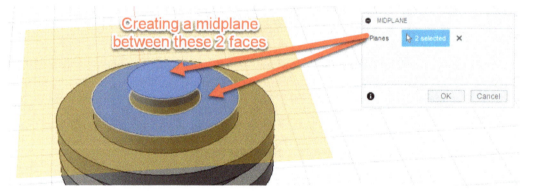

Figure 5.60 – The face selections for the Midplane tool

3.   Go to the **CREATE** dropdown and left-click on **Mirror**.

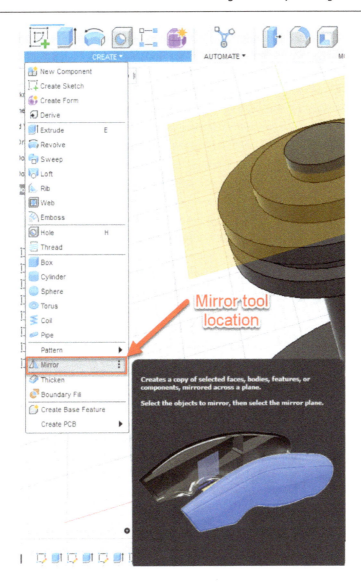

Figure 5.61 – The Mirror tool's location

4. Left-click on the **Object Type** dropdown and choose **Features**, then hold down the *Shift* key and left-click each of the extrusions that we previously created, as denoted by the red arrows in *Figure 5.62*:

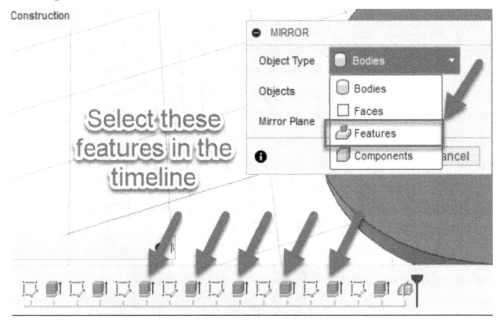

Figure 5.62 – The Features dropdown location within the Mirror tool

5. Now that the features that we want to mirror have been selected, choose the reference mirror plane that we created. A preview of the mirror will be shown.

Figure 5.63 – The mirror reference plane selection

6. Click on **OK** to close the **Mirror** tool.

7.   Left-click and turn off the Construction folder eyeball icon so that you can view the newly created doorknob.

Figure 5.64 – The final doorknob

The nice thing about creating the doorknob this way is that you can see which sketch corresponds to which part of the model and adjust just that part. Experiment with your model and see what interesting shapes you can come up with.

Summary

In this chapter, we went over three different ways to create 3D models: revolving by adding material, revolving by subtraction, and stacking by extrusion. There is no right way of creating a model, but ultimately it is up to you, the designer, to determine which way is the most efficient, and the fastest, to create the model, keeping the project budget and deadline in mind. In the next chapter, we will get a bit more creative and design a simple water bottle holder using calipers to take proper measurements on a bike frame.

Part 2:
Bicycle Water Bottle
Holder Project

In this part, we will begin a multi-chapter project building a water bottle holder for a bicycle. We will first learn how to sketch out some ideas on paper, use a caliper to collect measurements from a bicycle, create a reference model of the bicycle to gather measurements from, and then finally create the bottle holder. We will then improve upon our design and make more adjustments to the model, showcasing Fusion 360's parametric modeling power.

This part has the following chapters:

- *Chapter 6, Designing a Simple Bottle Holder*
- *Chapter 7, Creating a Bike Reference Model*
- *Chapter 8, Creating a Bottle Reference Model*
- *Chapter 9, Building the Bottle Holder*
- *Chapter 10, Improving the Bottle Holder Design*

6

Designing a Simple Bottle Holder

We are now about to start the first main project of the book, which is the design and modeling of a water bottle holder for a bicycle. Over the next few chapters, we will teach you how to take and record measurements from an existing bicycle using calipers and come up with a unique design that fits an 8-oz water bottle. In this chapter, we will learn how to create bottle holder sketches and use calipers to take measurements of the bicycle.

The first part of the project is to come up with a simple design and plan out how to approach the project. We will then come up with a few other design ideas and then add more details to the part. We will use some cheap calipers to determine the correct size of the bike support pipes and create the measurements of the bicycle holder according to the existing bicycle measurements taken. This will help you to learn how to take field measurements, adjust for proper fit, and determine the best design for what is required.

In this chapter, we're going to cover the following main topics:

- What is the project?
- Drawing out ideas
- Taking measurements with calipers

Technical requirements

You can practice with the files provided or feel free to create your own for a more customized experience. The sample design for this chapter can be found at https://github.com/PacktPublishing/Improving-CAD-Designs-with-Autodesk-Fusion-360/tree/main/Ch06.

Planning out the design

When working on projects, we need to first plan out the design, thinking about what exactly we want to accomplish. If you do a Google search for bicycle water bottle holders, you'll see a multitude of different ideas for various styles of bikes. We want to come up with a simply designed, 3D-printed water bottle holder that will hold a typical 8-oz plastic water bottle. It will need to be held in place on the bicycle pole with a grip or a clamp with a screw for tightening.

Now that we know what it is we want to accomplish, let's start to plan out the design.

I have taken a photo of the location where I want the bottle holder to be placed on my bicycle. For my bicycle, I want it on the pole underneath the handlebars, and I'd like to make a bottle holder for my wife's bike as well in the same location:

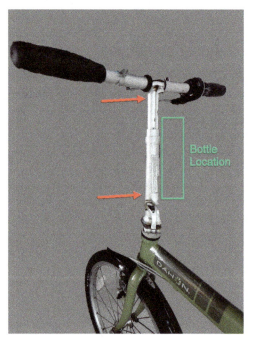

Figure 6.1 – My bicycle with a tapered pole

You can see via the photo of my bicycle that I have a tapered pole (see the red arrows), which means the pole starts out small and then gets larger as it gets to the bottom. This means that when I'm sketching out my ideas, I must consider that the lower bottle holder support is going to be larger than the top support as I will need two supports to hold the bottle holder in place.

For my wife's bicycle (see *Figure 6.2*), you can see that the pole is slightly different; it is not tapered so I will have to take into account that as well.

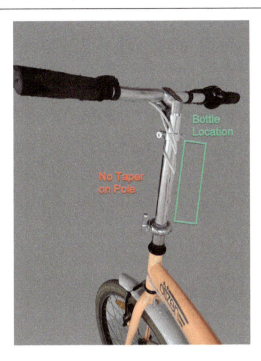

Figure 6.2 – An alternative style of bicycle with the same bottle location

When we create the water bottle holder, we will create two different versions of one design, with one having a taper and the other not. This will save on drawings and file space.

Drawing out ideas

Usually, when you want to design a part from nothing, you will need to sketch out some general ideas of what you want. Using a pencil and paper is still a good practice to have when thinking up ideas. I typically use butcher block paper since I can make it any size I'd like and have plenty of space to draw on for any other random ideas.

You may ask yourself though "*but I have this amazing Fusion 360 program to start my creation and draw in.*" Yes, you could just jump right into Fusion 360 and start building, but there are times when you can get lost in a project and build yourself into a corner and not know how to get out of it easily. You could also start making your design and then realize midway through that your approach may not have been ideal and you may run into problems with parametric dimensions.

I've drawn up some ideas of some general examples to follow but you can do this as well on your own if you have a great idea. Try to think about the general size of a water bottle, where it will be placed, such as in the front or underneath the bike seat, and what you would like it to look like. Come up with multiple versions so that you can spend time thinking about what you would like to create.

Version A

For this version, I would like the bottle holder to be as simple as possible. A good idea for starting is to just plan out the very simple basics that you are trying to accomplish. We need our bottle holder to be able to adjust to different-sized bottles and have a taper for the support holes. It needs support at the base and another support to keep the top of the bottle in place. See *Figure 6.3* for a simple idea.

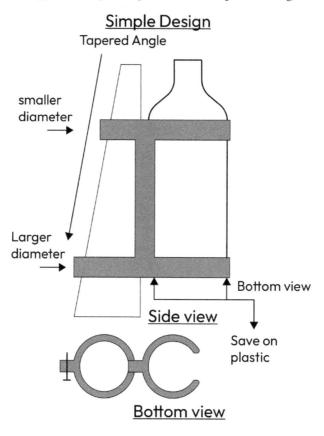

Figure 6.3 – Version A

In the following figure, I show the front, side, and top views to give myself a better idea of what I want to create:

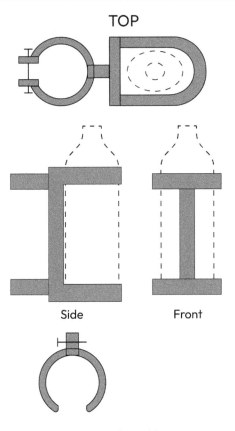

Figure 6.4 – Top, side, and front views

Creating this will really help when it comes time to work in Fusion 360, as you will get a better idea of the tools that you may want to use when creating the 3D model. You may start to see which tools we have used in the past few chapters to use here. Do you notice any way we can mirror a portion? Do you see any way that we can use the **Revolve** tool here? Is it better to use the **Extrude** tool? This is why simple sketches can help alleviate some pain when the time comes to create your object in 3D.

Version B

In this next version, let's step up the design a little bit and let's add some more flair to make it fun to look at. Get as creative as you'd like without thinking too much about the design aspect of it. Let your imagination run wild for this version.

As you can see in *Figure 6.5*, this version looks very interesting with a hook holder going part way around the bottle. I could also make this hook holder adjustable with a slide connection where it connects to the bicycle handle post. I like this idea and may hold onto it for the final version.

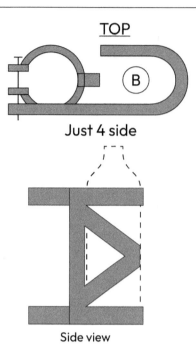

Figure 6.5 – A creative bicycle bottle holder design

You can see from this artistic version in the following figure that I have no way of making the bottle holder larger if I surround the entire piece with plastic.

Figure 6.6 – A spiral version of the bicycle bottle holder

I was looking to create a cool-looking spiral design but realized after looking at it that it may not work. This version looked great in my mind, but it does not meet the design intent of what we were originally going for. That's okay though; draw out as many ideas as possible so that when the time comes to create the 3D model, you will have a good idea of what will work and what may not.

Now, after some experimentation, let's try to combine both the simple version and the artistic version together and see what we can come up with.

Version C

In this next version, try to combine the best of both version A and version B, where a little flair meets good design. Usually, you will find that this version works out best as you get the best of both designs in one. Also, notice that I'm not trying to be neat at all when making these drawings. My handwriting has always been terrible, but I don't let that stop me when I'm creating. I know that when it comes to design and drawings in Fusion 360, the design will turn out to look beautiful in the end.

I like the idea of having two separate pieces that I can position if I have a taller or smaller bottle by just repositioning the bottle holder on the handle post, meaning that if I have a screw-tight grip on the bicycle handle post, I can loosen the screw on the bottle holder and move it up or down and then retighten the holder for larger bottles. I also think that the half-curved portion (see the top view of *Figure 6.5*) for the upper half of the bottle holder from the previous, version B, will work great for bottles of varying widths.

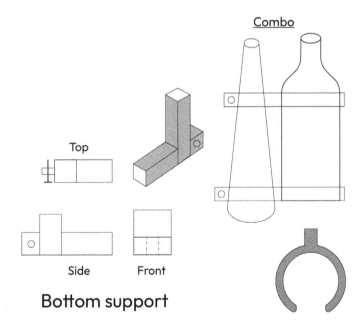

Figure 6.7 – Version C of bicycle bottle design

Notice that in all of the three designs that I did, I didn't add any dimensions just yet. I mainly just wanted to get an idea out first of what the bottle holder may look like. When we start using calipers in the next section, we will start to put together more of the design and dimensions.

Taking measurements with calipers

Now that we have a general idea of what we want to accomplish, it's time to take some measurements. If you already have calipers, then you are one step ahead of those of us who had to use a tape measure to get decent dimensions. For those who don't have calipers, you can get some cheap ones at any art store or online.

What are calipers?

Calipers are tools that measure the size of an object, typically of mechanical parts, with higher accuracy than tape measurement tools. There are two different versions of calipers – a digital one and a manual version (see *Figure 6.8*). I would suggest getting a digital one but if you're on a budget, a simple, manual two-dollar one will work just fine. I would also recommend getting a metric version since having a round number, rather than a number and a fraction, can make your life easier when taking measurements.

Figure 6.8 – A manual caliper

Calipers work by extending their pinchers and measuring either the outside diameter or, by using the indented pinchers, the inside diameter of a hole. You will notice on my small cheaper calipers (see *Figure 6.8*), the words **IN** and **OUT** for measuring the diameters of a hole.

In the next section, I'll use digital calipers that take measurements in millimeters and visually shows a digital measurement on a screen. As an American, it took some time to get used to the metric system, but now that I use it a lot more, I love having exact numbers rather than fractions.

Using calipers to take measurements

Now that we know what calipers are, how do we take good measurements the first time so that we don't have to go back and take more later? That is something that even most experts have tried to figure out for years. Just realize that sometimes, you can think and plan out everything, and still miss just one important dimension. Don't worry though; for our bicycle project, it should be relatively simple.

You will need to grab a notepad, a pen or pencil, the calipers, and your bicycle if you have one available. If not, feel free to use my dimensions to go by for this next exercise.

Draw out a basic shape of the bicycle water bottle holder location. I've taken a quick photo of my bicycle for reference:

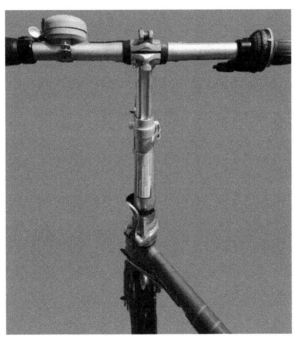

Figure 6.9 – Location where the water bottle holder will be attached to my bicycle

The shape doesn't need to contain everything yet; just get the basic height and width:

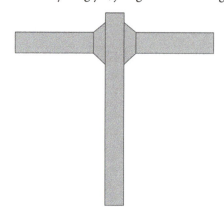

Figure 6.10 – Basic design of the bicycle handle post and handlebar

We will start to take measurements of the bicycle now. You will also need to grab a measuring tape if your calipers are not long enough to measure longer distances:

1. Measure the distance from the center of the top horizontal handlebar to the bottom of the vertical handle post (see *Figure 6.11* for reference). On my bicycle, that length is greater than my calipers, so I will need to use a measuring tape. The length comes out to be 270 mm:

Figure 6.11 – Tape measure distance to the center of the top horizontal handlebar

Now write that number down on your sketch. This is where you will need to start taking clear measurements because if your writing is bad, like mine, it will be hard for you to read them when you get back to your desk to put them into Fusion. Take your time writing these dimensions out clearly.

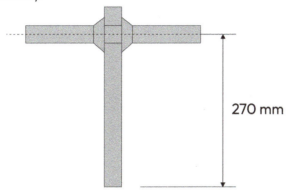

270 mm

Figure 6.12 – Bicycle handle post and handlebar sketch with overall dimensions

2. Since I have an adjustable bike handlebar, I will need to take a measurement to the top of the handle post clamp so that I can be aware of it when modeling in Fusion and will not accidentally hit it when creating the bottle holder. The measurement to the top of the bike handle post clamp is 163 mm.

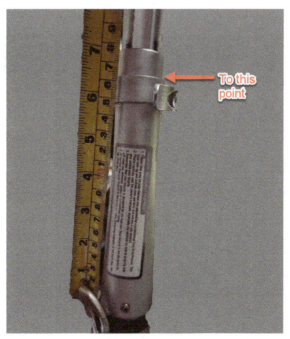

To this point

Figure 6.13 – Distance to the top of the adjustable bike handle post clamp support

Now write that measurement down as you did before.

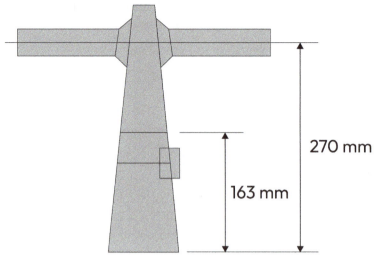

Figure 6.14 – Drafted-out measurement

3.   Now that we have the longest heights, we will now create the smaller measurements, which can be taken with some calipers so that we can get a much more accurate measurement. Starting from the bottom of the vertical pole, let's measure the diameter (see *Figure 6.15*). My measurement is 40 mm.

Figure 6.15 – Measurement of the bottom of the bike handle post using calipers

Be sure to write the measurement down very neatly, or close to it:

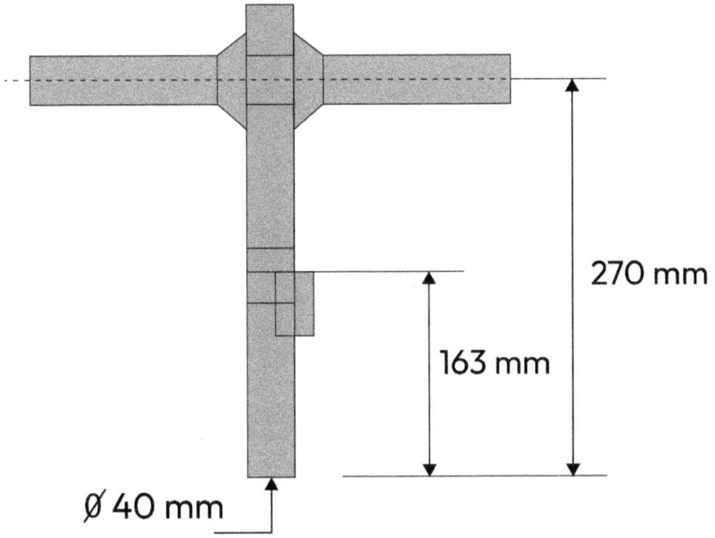

Figure 6.16 – Diameter dimension at the bottom

Keep taking measurements along the height of the pole and writing each one down neatly. If your bicycle has any features that may impact the location of the water bottle, be sure to take a clear measurement of their locations. We are making sure that the thickness and lengths are all accounted for in this area. For my bicycle, here are the measurements that I took, which we will use in the next chapter to create the bike handle post.

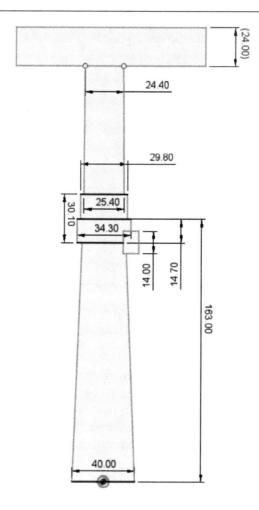

Figure 6.17 – All the dimensions for creating the bike handle post and handlebar

Now that we can get all the dimensions for the bike handlebar, we can start taking those dimensions and putting them into Fusion 360 so that we can create a reference model for our bicycle water bottle holder.

Summary

In this chapter, you learned how to plan and draw out ideas for a bicycle bottle holder to be placed on your own bike. You learned what the project is about, how to sketch out a few different designs, and how to take measurements using a caliper. In the next chapter, we will go over how to add in the dimensions that we took using the calipers to create nice-looking drawings that we can use for our Fusion 360 reference build.

7
Creating a Bike Reference Model

Now that we have created the pencil and paper notes for where we would like to place the bicycle water bottle holder, it's time to take those notes into Fusion and create a reference model of the bicycle frame. The reference model will be used not only as a general 3D model visual guide but also as a projected model that will be used as a reference for our 2D design sketches. For this chapter, we will only need the handlebars (top horizontal bicycle bar), the handlepost (vertical support bar), and the handlepost clamp, but if you would like to place your bottle holder elsewhere, such as under the seat, be sure that you have captured the necessary heights, widths, and thicknesses of the part as we did in *Chapter 6*.

By the end of this chapter, you will have learned how to take the dimensions from *Chapter 6* and create the bicycle parts that we will use as a reference model for the bottle holder. This bicycle reference model is important as we will use it to get a better idea of the size of the bottle holder we will create in the next chapter.

In this chapter, we're going to cover the following main topics:

- Planning the design intent
- Creating a reference model using primitives
- Creating a reference model using parametric designs

Technical requirements

You can practice with the files provided, or feel free to create your own for a more custom experience. The sample design for this chapter can be found at https://github.com/PacktPublishing/Improving-CAD-Designs-with-Autodesk-Fusion-360/tree/main/Ch07.

Planning the design intent

The first step we need to think about is the design intent so that we know what to expect out of this project and not stray from the path while we work. Since the bike's handlepost is a static object (meaning it's not going to change thickness), I could potentially create the bicycle handlepost with simple primitive shapes. For instance, I could use a simple cylinder for the handlepost and another cylinder for the handlebars. The only problem with that is my handlebars do change height, so I will probably need to use parametric objects so that they can change heights. I will show you both ways so that you can see the benefits of both.

Setting up the project

Grab the notepad that you used to write down the dimensions of the bicycle handlebars and handlepost as it's time to place those dimensions into Fusion 360. Here is the diagram from the previous chapter for reference.

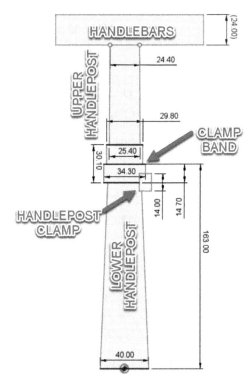

Figure 7.1 – Bicycle handlebar and handlepost dimensions

It's always a good habit to get into saving the project first; this way, you don't lose any data if the system crashes. Fusion does save your project while you are working, but if you never saved a drawing, then there would be no information to retrieve.

Here is how you can save your project in Fusion 360:

1. Open Fusion 360.

2. Click the **Save** button.

3. Create a folder within your Packt Publishing project and name it `Ch07 Handlebars Reference`.

4. Now name the Fusion 360 file `Bicycle Handlebars`.

Now that we have saved the project, let's change the units of the project to meters since the dimensions we took are all in millimeters:

1. Go to **BROWSER**, select the **Document Settings** drop-down arrow, and then mouse over the **Units** icon on the far right.

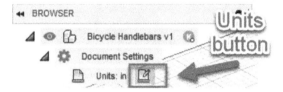

Figure 7.2 – Location of Document Settings

2. The **CHANGE ACTIVE UNITS** flyout will open. Left-click on **Unit Type** and select **Millimeter** from the drop-down list. Then click on **OK** to close the flyout.

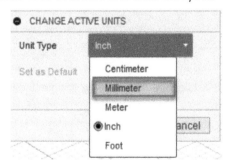

Figure 7.3 – The CHANGE ACTIVE UNITS flyout

Important note

You can still change between millimeters and imperial units just by typing in the unit of measure after a number. For instance, if I wanted to type in 6 inches even though my units are set to millimeters, I can type in `6in`. If I wanted to type in 6 millimeters even though my units are set to inches or feet, I could type in `6mm`.

Now that we have successfully changed the project units, let's create a new component by doing the following:

1. Within the **SOLID** tab, left-click on the **CREATE** drop-down arrow and select **New Component**.

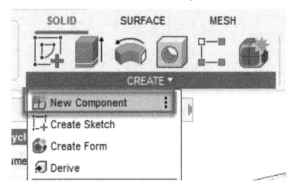

Figure 7.4 – Location of the New Component command

2. The **NEW COMPONENT** pop-up window will appear. Within the **Name** field, type in Handlebar Reference.

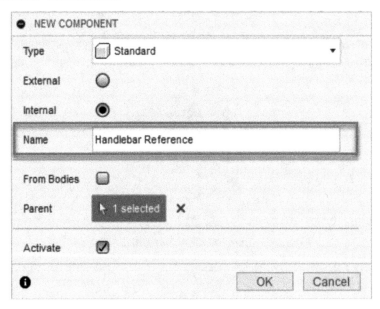

Figure 7.5 – The NEW COMPONENT flyout

A new component folder is created within the **BROWSER** on the left side of the screen, with the *Active dot* located at the right end.

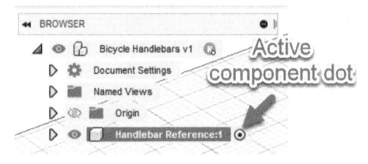

Figure 7.6 – The new component folder and active dot

> **Important note**
>
> Be sure that the *Active dot* stays on the **Handlebar Reference** component while you are working on this part. If you close out of Fusion 360 and get back into the file, it may move to the top of the browser, and any commands that you create will be made outside of the Handlebar Reference folder. Your Fusion 360 file will still work just fine, but it may not look clean and organized.

Now that the project setup is complete, let's first try creating the bicycle handlebar reference by using simple primitives.

Creating a reference model using primitives

Using primitives is a great way to get the basic shape of your reference model without having to use too many details. It's a quick and easy way to get your project moving forward. It is limited, though, by its basic function, so just be aware that this is not the preferred way to build since it is not a parametric model.

Here is how we can create our reference model of the handlepost and handlebars using primitives:

1. Within the **SOLID** tab, click on the **CREATE** drop-down arrow, and select **Cylinder**.

Figure 7.7 – The Cylinder tool location

2. Left-click on the *XY* plane and then left-click on the origin to place the center of the cylinder.

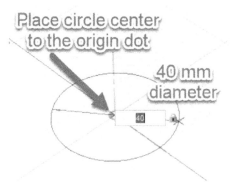

Figure 7.8 – Placing a cylinder on the XY plane

3. Type in a diameter of 40 mm and hit *Enter*.

4. The **CYLINDER** flyout will appear. Type in 270 for **Height** and hit *Enter*.

Figure 7.9 – The CYLINDER flyout

Now let's create the handlebar reference model:

1. Go to the **CREATE** drop-down arrow and select the **Cylinder** command once again. Left-click on the *YZ* plane this time.

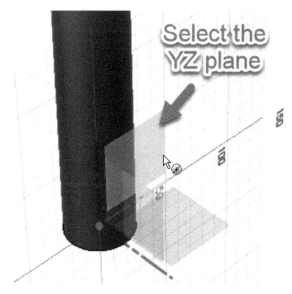

Figure 7.10 – The location of the YZ plane

2. Left-click on the **RIGHT** side of the ViewCube in the top-right corner of your screen. Your screen will now orientate itself from a perspective view to a right-side view.

Figure 7.11 – The RIGHT side of the ViewCube

3. The cylinder command should still be active even though you clicked on the ViewCube. Move your mouse toward the top center of the cylinder body that was just created and left-click when you see the center dot appear. Type in a diameter of 24.

24 mm diameter

Specify diameter of circle

Select the midpoint/center of the top cylinder

Figure 7.12 – The center dot location on the RIGHT side view of the previous cylinder

4. Type in 100 mm for **Height** (since we just need a general distance) and change **Operation** to **New Body**.

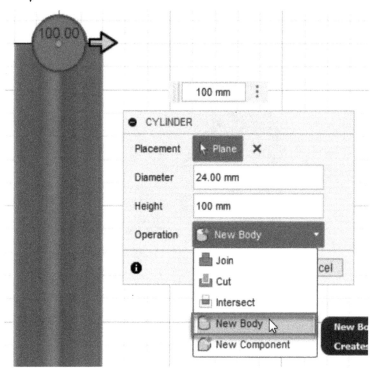

Figure 7.13 – The CYLINDER flyout window and Operation location

> **Important note**
>
> Notice that, by default, **Operation** is set to **Cut**. Since an intersecting object was created, Fusion thinks that you are creating an object to cut another object. This can be changed easily by going to the **Operation** section.

5. Click the *Home* icon within the View Cube in the top-right corner of your screen to set the screen back to a perspective view.

Figure 7.14 – Home icon within the ViewCube

6. Now let's mirror that part to the other side. Left-click on the **CREATE** dropdown and click on the **Mirror** tool.

Figure 7.15 – Location of the Mirror tool

7. The **MIRROR** pop-out window will appear. For the **Objects** selection, left-click on the cylinder that was just created, and then for the **Mirror Plane** selection, select the *YZ* plane. Leave **Operation** as **Join** and click **OK**.

Figure 7.16 – The MIRROR pop-out window and its options

The bicycle reference bodies have now been created. Notice that we did not have to create a sketch first to create these shapes. This is because it is not a parametric model but a model built from primitive, simple shapes. These work great if you just need a basic reference model to go by. You can still change the size of these shapes by going into the history bar at the bottom of the screen, but then they won't be connected to each other and will not move together. Creating it this way is still very limited due to the inability to quickly adjust the model by simply adjusting a single dimension, which doesn't make it ideal for future changes.

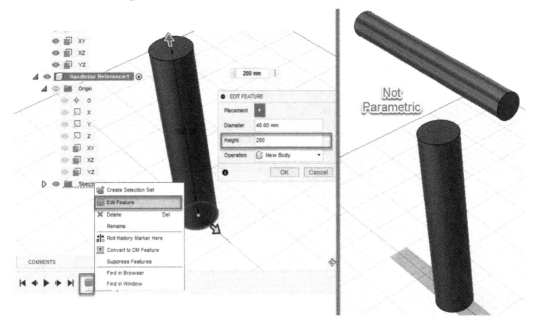

Figure 7.17 – A non-parametric model

Now let's create the same model but by using a parametric design instead. If a change is required, the rest of the model will flex and adjust accordingly depending on those changes.

Creating a reference model using parametric designs

We can see from the previous model that primitive modeling is a very crude, simple way of creating an object for quick reference. That method works great when you're short on time, but what if we need a few more details and want to see some size adjustments? This is where parametric modeling works very well. In this section, we will recreate the handlebars, the handlepost, and the handlepost clamp and demonstrate why parametric modeling is a much better way to build models:

1. With the top level active, right-click and select **New Component**. Name the new component `Parametric Handlebar Ref` and click **OK**.

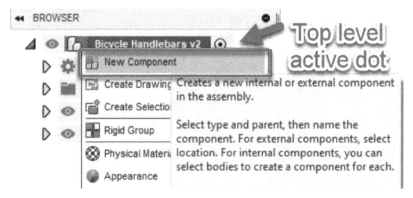

Figure 7.18 – Showing the top level active and New Component

Important note

If you happen to create the **New Component** within the previous component, your model will still work fine but may show incorrectly for a **Bill of Material** (**BOM**). If you notice that it was placed in the wrong location before you have modeled anything else, try to undo it and then continue modeling. If you have already created more models and do not want to undo all the way back, you can create a new component. Then, click and drag your model into that new component, but a moved icon will show up in your **TIMELINE** at the bottom of the screen. This is not a bad thing, but it is an extra added item in your timeline that could possibly mess up some edits in the future. It's usually best to keep your timeline as clean as possible.

---

2. Click the *Eyeball* icon of **Handlebar Reference** to hide the previous model.

3. Let's start from the bottom of the handlepost, as we did before. Click on **Create Sketch** and select the *XY* plane.

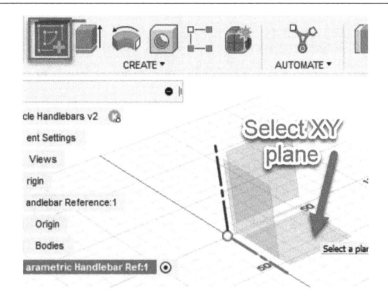

Figure 7.19 – Create Sketch tool location and XY plane

4. Click on the **Circle** tool and place the center point on the origin dot. Type in 40 mm for the diameter.

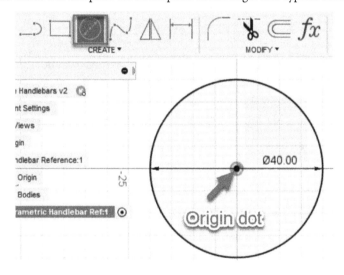

Figure 7.20 – Location of the Circle tool and origin placement

5. Click on **FINISH SKETCH** to exit the sketch environment.

Now we can extrude this circle, as we did before in *Figure 7.9*, but since we want to add a taper to this pole, we have two choices now. Within the **Extrude** tool, there is a taper option, but it is set to degrees. Since we do not know the taper angle, and took a physical dimension of the top diameter, we can use the **Loft** tool instead.

Using the Loft tool to create a taper angle

Using the **Loft** tool allows us to go from one shape to another shape. For instance, we could go from a square to a triangle or, in our case, a larger circle to a smaller one.

In this next section, we will use the **Loft** tool to create a slight taper angle for the handlepost model:

1. We will need to create another plane with a circle sketch to use the **Loft** tool. Select **CONSTRUCT** and then select **Offset Plane**.

Figure 7.21 – Location of the Offset Plane tool

2. Select the *XY* plane, set **Distance** to a height of 270 mm, and click **OK**. This will be the plane for the top center of the handlebars.

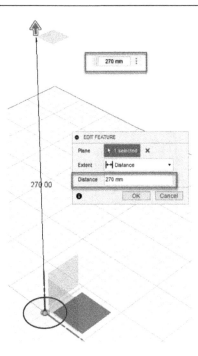

Figure 7.22 – Setting the height of the offset plane

3. We will need to do this for the other plane heights as well. Repeat this same process to create the top of the handlepost at 163 mm.

4. Create a sketch on the top plane at the center origin using the **Circle** tool with a diameter of 24.4 mm.

Figure 7.23 – Top sketch plane with a circle diameter

5. Now let's use the **Loft** tool to connect those two circles together to make a solid. Go to **CREATE**, then **Loft**.

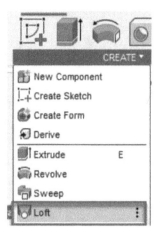

Figure 7.24 – Loft tool location

6. The **Loft** tool option popout will appear. Click on the bottom circle (40mm) and then left-click on the top. A preview of the **Loft** will appear on your screen. There are a lot of options here that let you adjust the curvature of the profiles and control the body shape with rails, but for now, click **OK**.

Figure 7.25 – Loft tool options

If your bike doesn't have an adjustable handle, you can stop here, but since my bicycle handlebar height can be adjusted and the adjustment handlepost clamp protrudes, I would like to add that to my reference model.

Adding the handlepost clamp

Now that we have the bicycle handlepost created, we will need to add in the handlepost clamp since this part protrudes where we want to place the bottle holder.

To create the handlepost clamp, we will create a sketch on the plane that is set at 163 mm and then extrude down to create the connection to the handle body, then add a sweep for the handlepost clamp.

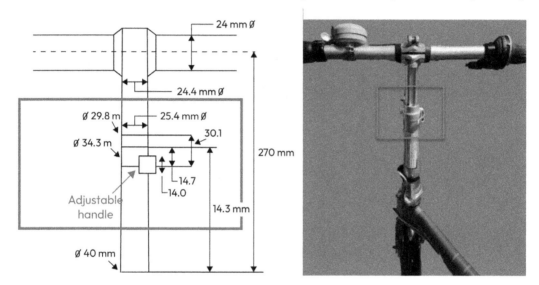

Figure 7.26 – Location of handlepost clamp

1. We will need to create a sketch on the 163mm plane first and then will use the body of the loft as a reference. Go to **CREATE**, then click on **Sketch**, and select the 163-mm plane.

Figure 7.27 – The 163-mm sketch plane

2. To create the profile of the circle at this location, we will need to use the **Intersect** tool. Go to **Project / Include | Intersect**.

Figure 7.28 – Project / Include | Intersect tool location

3. Once the tool is selected, the **INTERSECT** pop-up window will appear. Left-click on the bike frame loft body, and a red circle will appear showing the intersection of the offset plane that you created and the loft body:

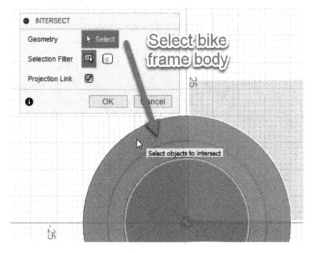

Figure 7.29 – The INTERSECT flyout selection

4. Click **OK** to finish the command. The red circle will now turn purple to show that it is a referenced circle.

5. Click on **Offset** within the **MODIFY** panel and set it to 3 mm.

6. Click on **FINISH SKETCH** to go back to the **DESIGN** workspace.

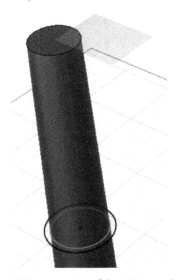

Figure 7.30 – Location of the 163-mm sketch

If we use the **Extrude** option right now, it will cut into the lofted body, due to the taper angle. We need to create another offset plane 14.7 mm down, then create another sketch with an intersection, and then loft to that intersected sketch.

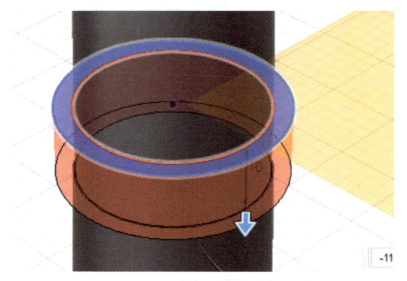

Figure 7.31 – Extruded cut into the lofted body

We could create the extrusion with a draft angle, but since we are not sure of the draft angle, we will create another loft. To do this, we must first create another sketch a few millimeters down, and then we will project the existing geometry of the handlepost and offset to create another circle. Then, finally, loft between the two sketches to create the handlepost band:

1. Go to **CONSTRUCT**, then **Offset Plane**, and select the plane at 163 mm to offset from. Select the directional arrow and pull it down to set it at a height of -14.7 mm, and select **OK**.

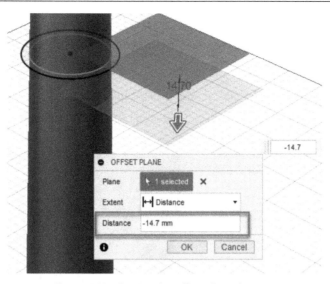

Figure 7.32 – Setting the offset plane distance

We are selecting this plane and not the bottom origin plane because if the 163-mm height should change, this height will move with it because we are telling Fusion to always stay -14.7 mm away from the 163-mm plane.

2. Go to **SOLID**, then **Create Sketch,** and select the new –14.7-mm plane that was just created.

3. Select the **CREATE** drop-down arrow, then go to **Project / Include,** and then select **Intersect**, as you did previously. Select the handlepost body rather than the outer line. If it becomes difficult to select objects, use **Selection Filter** and set it to **Body**. Click **OK**.

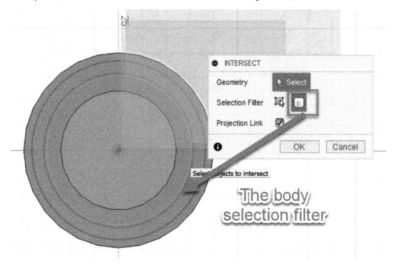

Figure 7.33 – Selection Filter of the Intersect tool

4. We need to offset this circle, but it may be hard to see due to the other circle above. Go to **BROWSER** on the left side of your screen and open the `Sketches` folder. Turn off all other sketches except for the one you are in so that you can clearly see the circle. Now go to **MODIFY**, then **Offset** or hit *O* for the keyboard shortcut, and select the circle, following which we set the offset distance to 3 mm. Click **OK** to finish the **Offset** command.

5. Click **FINISH SKETCH** to exit the **Sketch** environment. In the **BROWSER** window to the left, turn on the sketches we turned off in *step 4*.

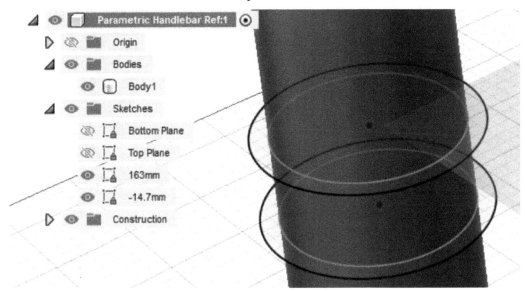

Figure 7.34 – Sketches that are ready for the next loft

6. Go to the **CREATE** drop-down arrow, the **Loft** tool, then select the `163mm` and then the `-14.7mm` sketch profiles. Notice that the **Loft** tool selection defaults to the **Cut** operation since it thinks that you want to use the two sketches to cut a section out of the other body. Click on the **Operation** option near the bottom of the **LOFT** options pop-up window and change it to **New Body**. Then, click **OK**.

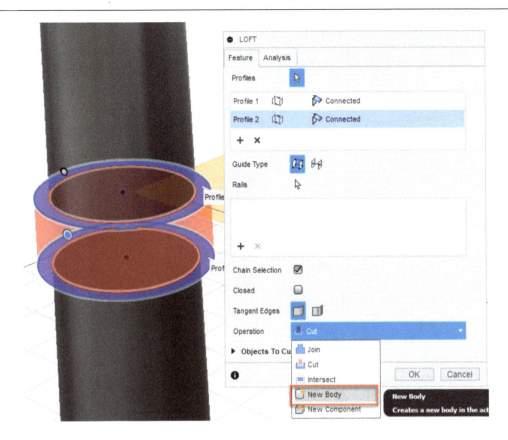

Figure 7.35 – The Loft tool's New Body operation

Important note

If you don't like that Fusion 360 defaults to the **Cut** operation, you could also turn off the *Eyeball* icon for the bicycle support bar body, and Fusion will default to create a new body since it will no longer "see" the other body.

If you hide the bicycle support bar body, you will notice that there is no hole for the loft.

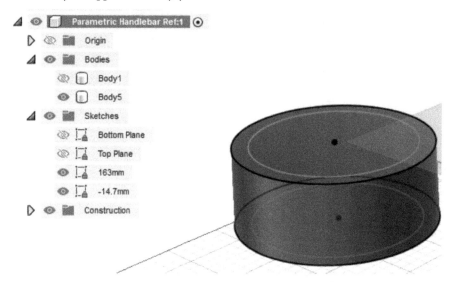

Figure 7.36 – No hole for the lofted body

This will happen for any lofted sketch as the **Loft** tool will choose the entire sketch plane as the shape. To create the hole cutout, we can use the **Shell** tool.

Creating a hole cutout with the Shell tool

The **Shell** tool is a handy command that will add a hollow void within a 3D body. You can also remove singular faces, multiple faces, or none to create an empty shelled object. We will be using this tool to create a void inside the handlepost band:

1. Go to the **MODIFY** panel dropdown and select the **Shell** tool.

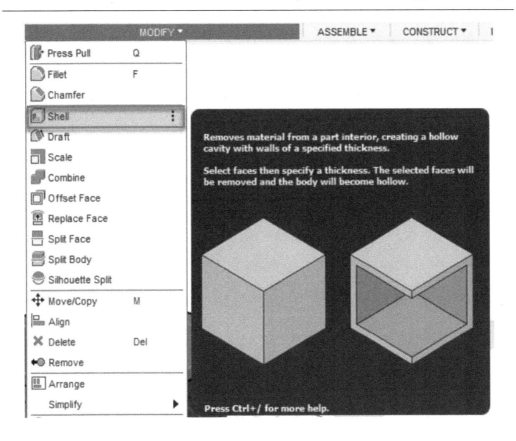

Figure 7.37 – Showing the Shell tool

2. Now select the top face of the loft that we just created, then hold down your right mouse button, and drag it forward to orbit to see the bottom face (you could similarly left-click on the View Cube's bottom corner). Now, left-click on that face as well. We select both faces since we want to remove both the top and bottom faces for the shell. Set **Inside Thickness** to 3 mm and click **OK**.

Figure 7.38 – The Shell tool options

> **Important note**
>
> You may have noticed that the tool name within *Figure 7.38* says **EDIT FEATURE**. This is still the **Shell** tool, but the **EDIT FEATURE** name appears when you go back to fix the tool from the timeline. The options within the command don't change from the original, just the name.

3. Click on the *Eyeball* icon to turn on the handlepost body that was previously turned off. Then, turn off the 163mm sketch and the -14.7mm sketch.

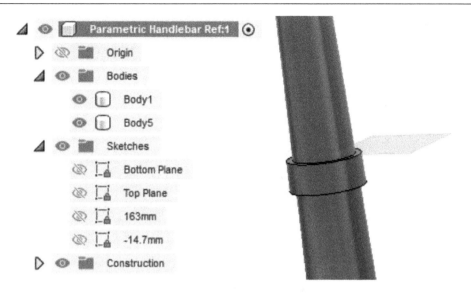

Figure 7.39 – Sketches off and bodies turned on

The next part is to add the handlepost clamp as a reference. You will notice that the clamp has a lot of curves to it, but we are going to just show this part very simply. If this part broke and we needed to model a new handle, then we would use our calipers to add more details, but since we just need to see this part as a location reference, we can keep it simple. This is typical for highly detailed parts that are being used for reference. Reducing the amount of geometry will save on computer resources and keep your file size small and easily workable.

Creating the reference handlepost clamp

The handlepost clamp sticks out a little bit from the handlepost, so we will need to show that in our model so that when we create our bottle holder, we know where it is placed.

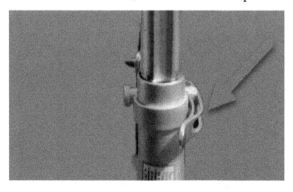

Figure 7.40 – The adjustment handle

To create this handlepost clamp, we will use another loft, but this time, we will manipulate the geometry so that it adjusts its curvature to go around the handlepost. In this section, we will create the first loft sketch:

1. Click on **Create Sketch** and left-click on the *XZ* plane. Your screen will display the front view.

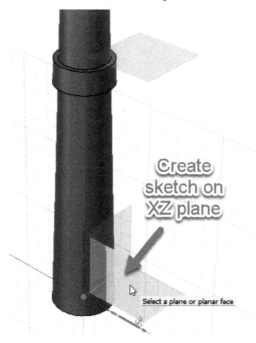

Figure 7.41 – Location of the XZ plane

2. Zoom into the handlepost band from *Figure 7.39* and then go to the **CREATE** drop-down arrow, then **Project / Include**, and select **Intersect**. Now, left-click on the handlepost band and click **OK**.

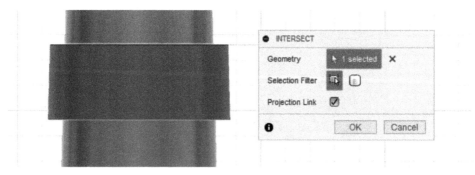

Figure 7.42 – The INTERSECT tool

3. The purple lines along the edges are projected from the 3D model of the handlepost band. This means that if that model changes shape, size, or location, then it will also be adjusted here as well. Use the **Line** *tool* (*L* keyboard shortcut) and draw a shape like in the following screenshot, as this will represent the rear, thicker part of the handlepost clamp. Be sure that the endpoints show up as black dots, which means that they are constrained/connected to the model. If they show up as white, then you will need to use the *Coincident* constraint to connect them back together again.

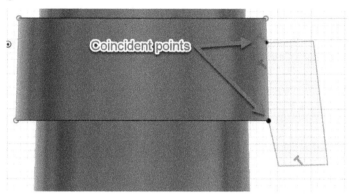

Figure 7.43 – The 2D line sketch

4. Now let's add some more constraints to the 2D sketch. Select the *Parallel* constraint and select the purple line, then the outer line (red arrows). Now select the *Colinear* constraint and select the purple line underneath to make them on the same linear path (green arrows).

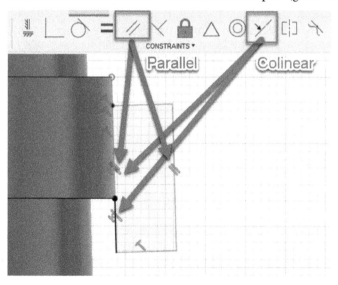

Figure 7.44 – Additional constraints added

5. Now let's set the dimensions and make the sketch fully constrained. See the following figure. Press the *D* keyboard shortcut, select the two horizontal lines (red arrows), and set the distance to 14 mm. Now select the purple dot and the top corner point (green arrows) and create the 3-mm dimension. Then select the same points for the 7-mm dimension. Click on **FINISH SKETCH**.

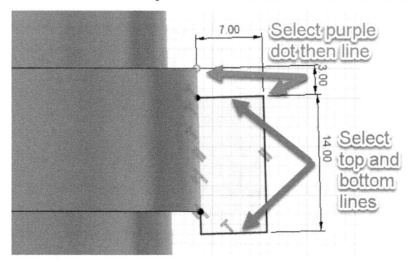

Figure 7.45 – Location of the dimensions

> **Important note**
>
> A "fully constrained" or "fully defined" sketch means that all the sketch line objects are now only controlled by the constraints and dimensions that are connected to them. All the lines will turn black, letting you know that the sketch has been "fully defined." When you create the 3D model from the sketch, it should change shape without any issue due to the sketch being "fully defined." If you notice that a line or circle is still blue, this means that it could possibly still move. You may need to constrain it with a horizontal or vertical dimension or with one of the many constraints, such as coincident, horizontal/vertical, and so on.

We could simply extrude this shape, but since it curves inward, we will create another, sketch out the length of the handle, and then loft to that sketch.

Creating the second loft sketch

Now that we have the thicker end of the handlepost clamp sketch created, we will create the thinner end next. First, we will create an offset plane, then make a smaller sketch on that plane, which we will then use to loft the thicker and thinner planes together:

1. We need to create the offset plane first, so go to **CONSTRUCT**, then click on **Offset Plane**, then left-click on the sketch we created previously, set **Distance** to -32mm, and click **OK**.

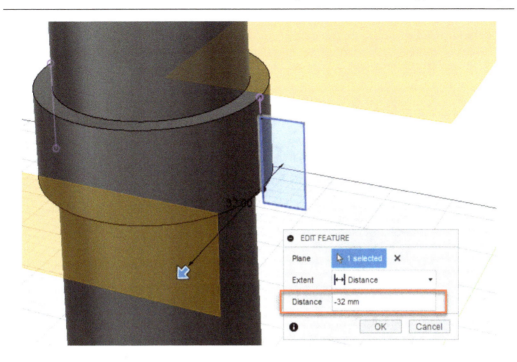

Figure 7.46 – Setting the offset plane distance

2. Go to **CREATE**, **Create Sketch**, and select the offset plane we just created. Now press the L keyboard shortcut and then the X keyboard shortcut to set the line to a construction line, as seen in the following figure. Left-click on the top middle of the cylinder body and the bottom middle to set a midpoint constraint to the center (red arrows). Now create two construction lines (orange arrows) from the previous sketch. Now hit the X keyboard shortcut again to turn off the construction line option and create a solid line (green arrow).

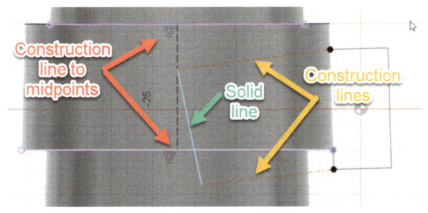

Figure 7.47 – Construction sketch lines

3. Now we set the constraints for this sketch. Set collinear constraints for the horizontal lines (orange arrows in the following figure) and set a parallel constraint for the vertical solid line (red arrows in the following figure). Notice everything turns black, so we will not need to set dimensions.

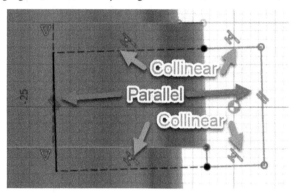

Figure 7.48 – Added constraints

4. Set an offset for the solid line of 1 mm.

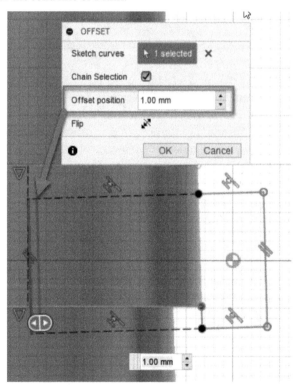

Figure 7.49 – Setting the offset distance

5. Start the line tool (*L* keyboard shortcut) and add two lines at the top and bottom (red arrows in the following figure). This will complete the profile so that it is a selectable face. We need to add these two lines since the construction lines do not show up to generate a profile sketch. Notice that the purple reference lines split the profile into two selectable faces. It will be easier when selecting a face to turn these purple reference lines into construction lines. Select them both and hit the *X* keyboard shortcut (orange arrows in the following figure). Click **FINISH SKETCH**.

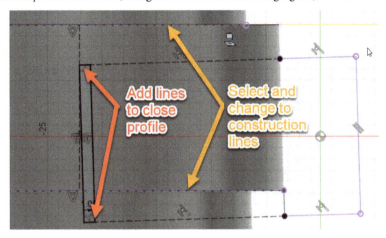

Figure 7.50 – The added solid lines and purple reference lines

6. Start the **Loft** tool and select both profiles. Notice that it defaults to the **Cut** operation due to the direction of the profiles.

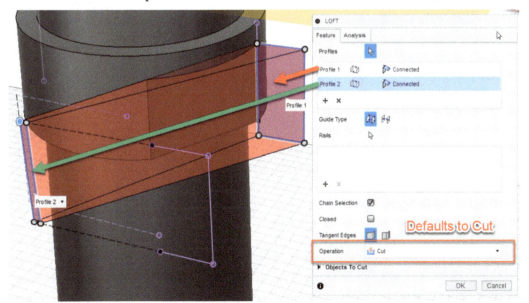

Figure 7.51 – The LOFT options flyout

7. We could either create another sketch and create a rail path to tell the loft to go around the body or we could select the word **Connected** on **Profile 1** and change the dropdown to **Direction**. Notice now that it goes around the body, due to the location of the profiles and **Takeoff Weight** and **Angle**. Experiment with some of these numbers and see the various design changes. When you're finished, set the options back to where they were, set **Operation** to **Join**, and click **OK** to finish the command.

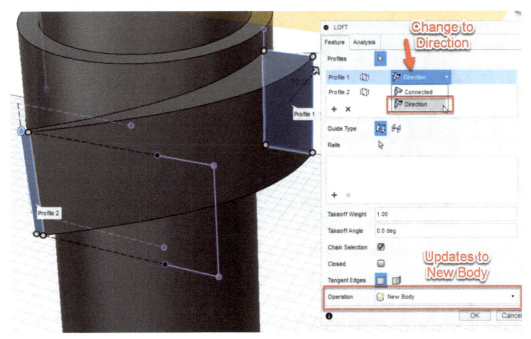

Figure 7.52 – Showing the loft tool's Direction option

Now that we have finished the handlepost and clamp, we just need to add in the handlebars, and our bicycle reference model will be completed.

Figure 7.53 – Finished handlepost model

Creating the bicycle handlebars

The handlebars can be created in a few different ways. Which way would you like to create them based on what we learned in *Chapter 5*? Should we create a profile sketch and revolve it? Should we extrude it in multiple sections? Hopefully, you're starting to see how we can generate objects just by looking at them. We're going to keep it simple again, as this model is basically a reference design model.

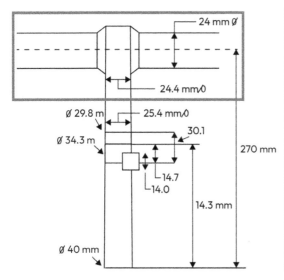

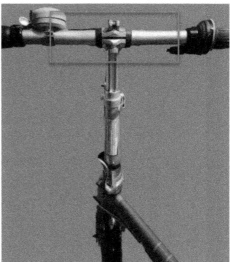

Figure 7.54 – Handlebar image and dimensions

Let's make our lives easy and create a simple revolve by starting from the middle. Be sure to stay within the **Parametric Handlebar Ref** component; otherwise, your sketches and bodies will be created outside of the component, which will make it hard to follow for other designers.

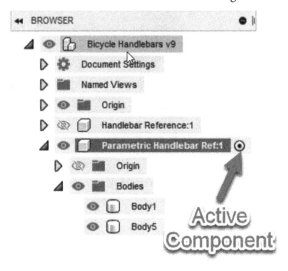

Figure 7.55 – Active dot on the Parametric Handlebar Ref container

Before we move on, though, there is a small error in the model. When I created this model, I mistakenly thought that the 163-mm height was to the top of the lower handlepost, but it was supposed to be the top of the handlepost clamp band. No problem – with Fusion 360, we can change this easily and keep moving forward.

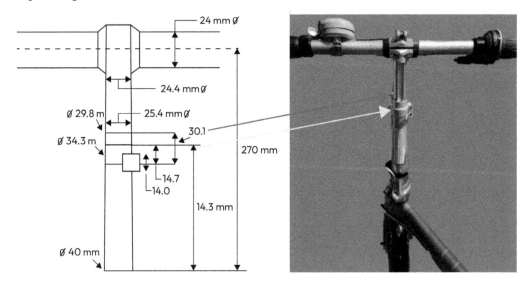

Figure 7.56 – Showing the change that needs to be made

Fixing the support height

If you notice in my dimensions, the 163 mm goes to the top of the handle adjustment, and then I measured down 14.7 to the bottom. I then measured from the bottom of the adjustment handle up 30.1 to where it meets the handlebar slider. So, to fix my mistake easily, I will move the reference plane down from 270 mm to where it should be, then create an extrusion up to 270 mm.

We need to do a little math first. We need to find the difference between 30.1 and 14.7, which is 15.4. Now we need to add that to 163, which is 178.4 mm, which is our new height from the bottom of the lower handlepost to the top of the lower handlepost:

1. Go to the timeline and mouse over some of the icons at the bottom. Notice that when you move over some icons, they will highlight the object within your screen. This is an easy way to tell which icon relates to which object. Right-click on the second plane object and select **Edit Feature**.

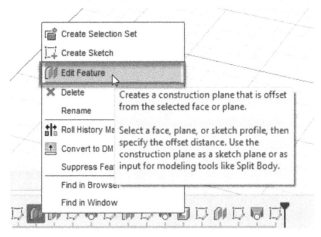

Figure 7.57 – The Edit Feature command within the timeline

2. Once you edit the feature, you will notice that the timeline fades and all your 3D objects disappear. This is because you are essentially going back in time back to when you first created that feature. Change the height (**Distance**) to 178.4 and click **OK**. Your model will adjust to the new height.

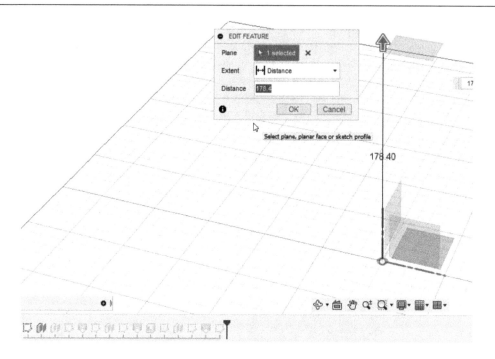

Figure 7.58 – Changing the height within Edit Feature

3. Go back into the timeline at the bottom of your screen, right-click on the second sketch created (this should be the top of the lower handlepost), and select **Edit Sketch**. Change this circle to 29.8mm and click on **FINISH SKETCH**.

Figure 7.59 – The Edit Sketch location

4. Go to **Create Sketch** and select the top face of the bicycle frame body.

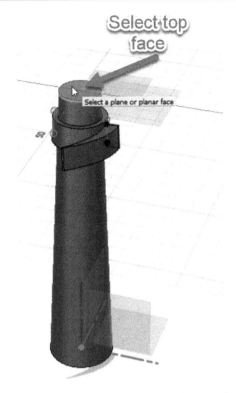

Figure 7.60 – Showing the selection of the top face

5. Create a circle connected to the origin dot and set the diameter to 25.4mm. Click on **FINISH SKETCH**.

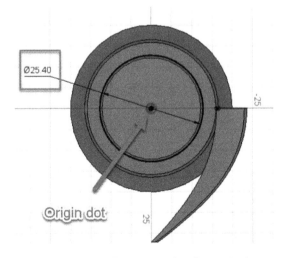

Figure 7.61 – Circle connected to the origin dot

6. Create an offset plane by going into the **CONSTRUCT** panel and setting **Distance** to 91.6. This number is the original height of 270 mm - 178.4 mm.

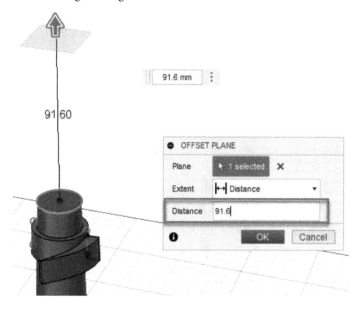

Figure 7.62 – The offset plane Distance

7. Create a sketch on this plane and then create a circle with a diameter of 24.4 mm. Then click on **FINISH SKETCH**.

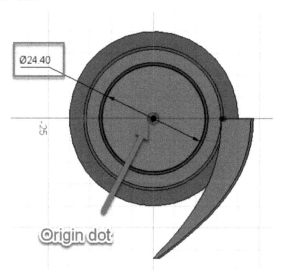

Figure 7.63 – Showing the next circle

8. Turn off **Body 1** to better see the lower sketch profile by clicking on the *Eyeball* icon in **BROWSER**. Start the **Loft** tool and select the two profiles and click **OK**.

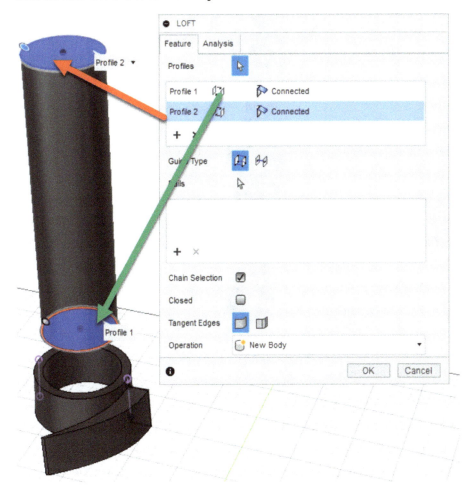

Figure 7.64 – The loft between the two planes and Body 1 off

9. Turn **Body 1** back on to see the finished adjusted body.

Figure 7.65 – The finished bicycle body

Now this resembles the image a bit better. Let's finalize this reference model by creating the handlebars.

Creating the handlebars

Now we will finalize the bicycle reference model by adding in the handlebars. The easiest way to do this would be to create a sketch profile, then use the **Revolve** tool along the center axis to create the 3D model. We will then use the **Mirror** tool to flip that handle to the other side to finish the handlebars:

1. Start the **Create Sketch** tool with the **CREATE** panel and select the *XZ* panel.

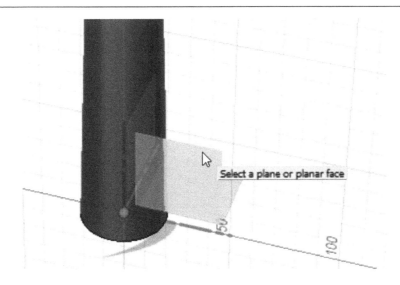

Figure 7.66 – The XZ panel

2. The display will change to the front view. Create a simple line design (as shown in the figure) resembling half of the handlebar since we will do a revolve.

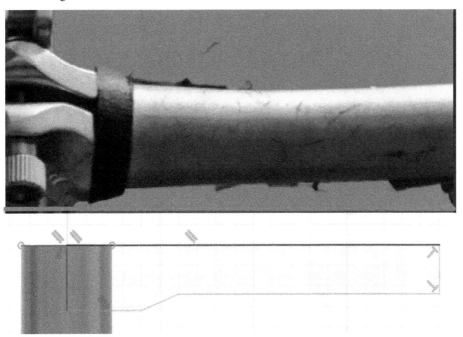

Figure 7.67 – Handlebar line design

3.   Add the dimensions shown in the figure for the lines and click **FINISH SKETCH**.

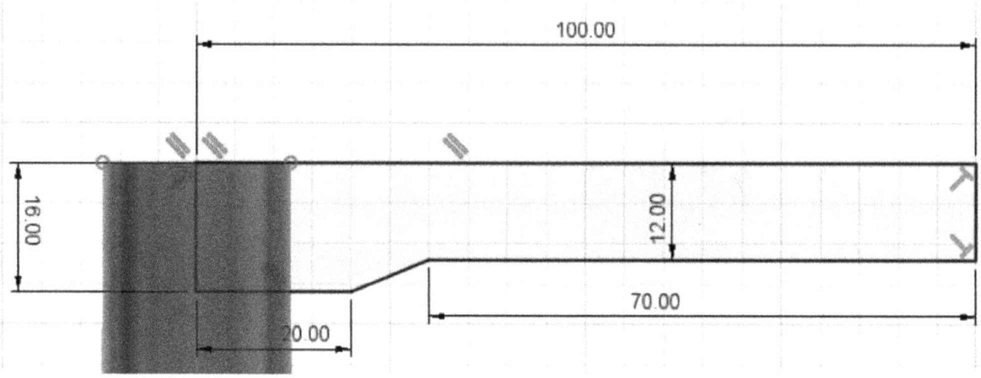

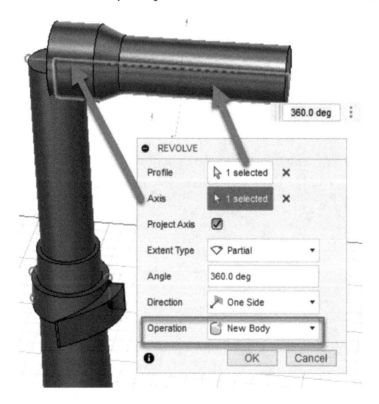

Figure 7.68 – The fully constrained sketch

4.   Click on the **Revolve** tool located within the **CREATE** panel and select the sketch that we just created if it wasn't selected already. Select the top line (the red arrow in the following figure) for **Axis** and select **New Body** for **Operation**. Click **OK** to finish the command.

Figure 7.69 – The Revolve tool options

5. Select the **Mirror** tool within the **CREATE** drop-down arrow and then select the revolved body we just created. Select the flat face at the end as the mirror plane. Click **OK** to finish the command.

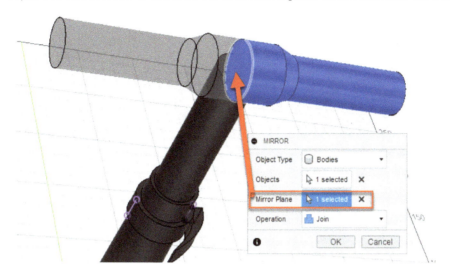

Figure 7.70 – The Mirror tool

Congratulations! We now have our finished reference model.

Figure 7.71 – The finished bicycle handlepost and handlebar reference model

I hope that you can create your own bicycle handlepost and handlebar reference model as well using the techniques that we went over. Don't worry about making mistakes; these will happen when you are first learning, but once you understand how to fix them, you'll feel much more confident when you make your next model.

Summary

In this chapter, we learned how to create a bicycle reference model with two different techniques: one using 3D primitives and one using parametric designs. We first learned how to plan out the design and then created basic models with the aforementioned techniques, which we can use for other similar bicycles. It is up to you, the designer, to know what may be best for building your model. Will you need every detail shown or just the basics to give you a general idea of where parts sit? Knowing the difference will help you become a much more efficient designer.

In the next chapter, we will create a water bottle reference model, which we will use for our water bottle design.

8

Creating the Bottle Reference Model

In this chapter, we will use the calipers from *Chapter 6* and take the dimensions of a water bottle to create a reference model for our water bottle holder, which we will create in the next chapter. The water bottle that we create will be used to manipulate the size of the bottle holder so that if a change is needed, we can adjust the size of the bottle to adjust the size of the holder. This chapter is mainly about creating a 3D model, which does not contain all the fully detailed features such as bottle-cap threads or a hollow bottle. We will learn more about creating parametrics and using those parametrics to manipulate the shape of the bottle body.

We will learn how to model a bottle from the dimensions measured with our calipers and then put those dimensions into Fusion 360. We will then use the **Change Parameters** tool to adjust those dimensions to create the other bottle sizes.

In this chapter, we're going to cover the following main topics:

- Measuring the dimensions of a water bottle
- Creating a 2D parametric reference model
- Creating a 3D reference model
- Referencing both models together

Technical requirements

You can practice with the files provided or feel free to create your own for a more customized experience. The design example for this chapter can be found at https://github.com/PacktPublishing/Improving-CAD-Designs-with-Autodesk-Fusion-360/tree/main/Ch08.

Measuring the dimensions of a water bottle

As we learned from *Chapter 6*, calipers are a helpful tool for capturing the dimensions of real-world objects. In this section, we will capture the dimensions of a few water bottles of different sizes: a large pink metal bottle, a medium-sized red metal bottle, and a small plastic bottle. We will measure their dimensions and note them down so that we can bring those dimensions into Fusion and find a general size for the bottle holder. All these bottles maintain an almost constant diameter throughout their length, so we won't have to worry about varying widths too much.

Figure 8.1 – Various-sized water bottles

Let's look at typical water bottle sizes and get the dimensions for them. As you did before, grab your calipers and a measuring tape to capture the overall length and width. This time, I won't go over each step but will show the dimensions that I measured. If you have a bottle, I encourage you to practice taking dimensions yourself and see how close your dimensions are to mine. Once again, I took all my dimensions in millimeters as it is easier to work with whole numbers, rather than fractions.

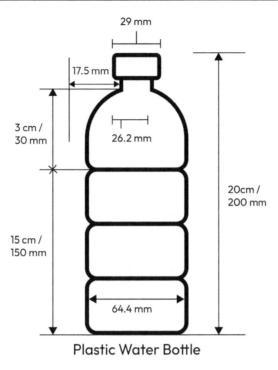

Plastic Water Bottle

Figure 8.2 – Dimensions of a typical small plastic bottle

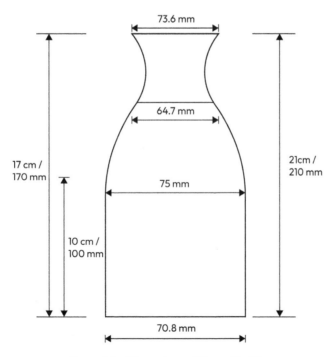

Figure 8.3 – Dimensions of the red bottle

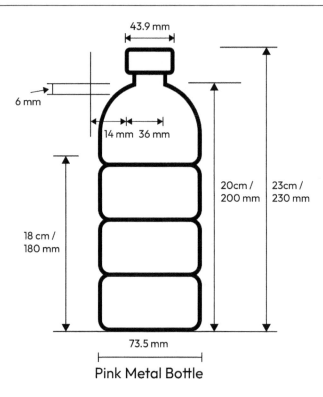

43.9 mm

6 mm

14 mm 36 mm

20cm /   23cm /
200 mm   230 mm

18 cm /
180 mm

73.5 mm

Pink Metal Bottle

Figure 8.4 – Dimensions of the pink bottle

It's OK to round off the dimensions as well. As you can see, some of my dimensions have decimal places but since these are being used for reference once again, we could round off all the dimensions to get nice even numbers once we are in Fusion 360. I would recommend though that you try to take dimensions as closely and accurately as possible while you have the physical object to hand, as sometimes the more precise dimension may come in handy.

If you notice, in all the three water bottles that I measured, I wasn't sure what the radius was for the curved top edge. This curved area is what is known as a fillet. A fillet is a nice smooth curve and is mainly placed on corners so that people do not cut their fingers on sharp corner points. Since I wasn't sure of that curvature, to capture that radius, I chose two endpoints where the curve starts and stops, then took the length (the x axis) and height (the y axis) dimensions. I then will then insert those dimensions into Fusion and figure out what the curve is between those two end points.

Now that we have all our dimensions, we can take an average dimension from them and use that for our water bottle reference model. We now know that our water bottle holder must be able to hold a water bottle that has a diameter of between 64 mm and 73 mm and a height of between 20 cm and 23 cm. The numbers, averaged out, come out to around to the height and width of the middle boddle so we will use that for our reference size.

Creating a 2D parametric reference model

Let's take the middle-sized water bottle and use that for our reference. Remember, since Fusion 360 is a parametric model, we can easily create the other models just by changing the dimensions. This time, we will use the parametric manager and use labels such as length, width, and height to change the size of the bottle.

Starting a new project

Open Fusion 360, hit the **Save** button, and create a folder within your PACKT Publishing project, naming it Ch08 Bottle Reference. Now name the Fusion 360 file Bottle. Change the **Units** to **Millimeter** since we took all dimensions in millimeters (refer to *Chapter 7* in order to see how to change the units if required).

We can either start the sketches, add the dimensions, and then use the **Change Parameters** manager to add the parametric label names, or we start off by adding the the **Length**, **Width**, and **Height** parametric names first since we know what names we want to use. Either way, you can always go into the manager and add the names whenever you'd like to.

The Change Parameters manager

To get to the **Change Parameters** manager, you can either click on the *fx* logo within the **MODIFY** panel or go to the **MODIFY** panel drop-down arrow and left-click on the **Change Parameters** option to open the manager popup. If the *fx* icon isn't in your top panel, open the dropdown and mouse over the far-right corner until you see three dots appear. Left-click on the dots and you can pin the icon to the top. This works for all icons.

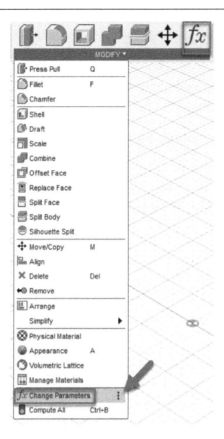

Figure 8.5 – Location of the Change Parameters tool

Once you click on the icon, the **PARAMETERS** pop-up window will appear. Since we haven't created any sketches with dimensions yet, no information is listed within the **Model Parameters** area, but we can add in **User Parameters** such as the length, width, and height plus any other features that may be a part of the model.

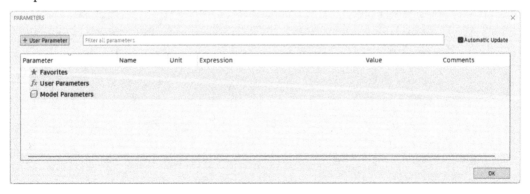

Figure 8.6 – The parameters pop-up window

Let's add some user parameters now:

1. Click on the **+ User Parameter** button in the top-left corner, which should open the **Add User Parameter** pop-up window.

Figure 8.7 – Add User Parameter pop-up window

2. In the **Name** field, type in overallHeight. In the **Expression** field, type in 210. Notice that you can change the units by using the dropdown or by typing in the units with the number, e.g., 210 mm in the **Expression** box. Click **OK** to finish. The user parameter is added to the **User Parameters** field.

Important tip

There are no spaces allowed in the **Name** field. If you want to add a long name that would normally include spaces, I usually employ what is known in the coding world as "camelCase text", meaning a lowercase letter starts the first word, and an uppercase letter starts the following word(s) e.g., overallHeight.

We can add math expressions into the **Expression** field as well. Let's add the bottom diameter dimension.

3. Click back on the **+ User Parameter** button and add a name of bottomDiameter and an expression of 70.8/2. The number 70.8 is the full diameter but we can use a formula and divide by 2 to get the radius dimension. We are using a radial dimension since we will create half of the bottle sketch using a construciton line down the middle and use the **Revolve** tool to create a solid body.

4. Add in the rest of the user parameters as shown in *Figure 8.8* and click **OK** when you are finished. If you make any mistakes, you can always edit the text or the dimension number by left-clicking on the text. At this time, you cannot reorder the parameters, but it honestly doesn't matter since we will reference them in a sketch.

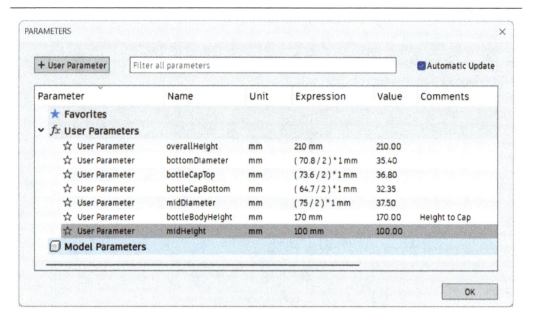

Figure 8.8 – User parameters for the water bottle sketch

Now let's generate the sketch to go with the user parameters that we just created.

Creating the sketch

The sketch we will create will only be half of the bottle body since we will be using the **Revolve** tool to generate the model:

1.   Click on the **Create Sketch** button located within the **CREATE** panel, then click on the *XZ* plane.

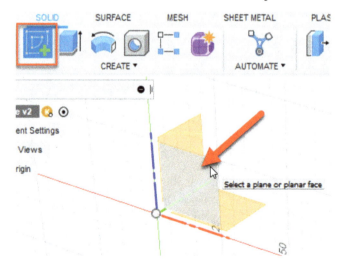

Figure 8.9 – Location of the Create Sketch tool and the XZ plane

2. Create a sketch that is roughly the shape shown in *Figure 8.10*. Do not worry about the dimensions or size just yet. We will just create the general shape we're looking for in Fusion and then add in the parameters and dimensions.

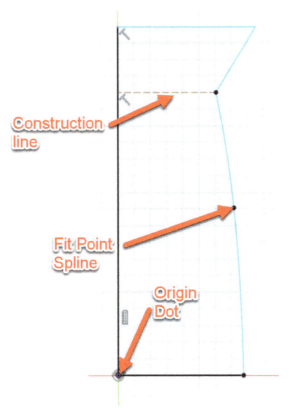

Figure 8.10 – Sketch the general shape without dimensions

3. Now we can add the dimensions with the user parameters that we created earlier. Usually, I start with the overall height of the sketch since this will grow the sketch to the size that we want. Press the keyboard shortcut *D*, left-click on the construction line, and then left-click away from the line to place the location of the dimensions. The dimensions will be highlighted in blue. Type in the letter o and notice that once you type in that letter, you are provided with a list of parameters with the letter o within them. Left-click on **overallHeight** from the list and then hit *Enter* to finish the command.

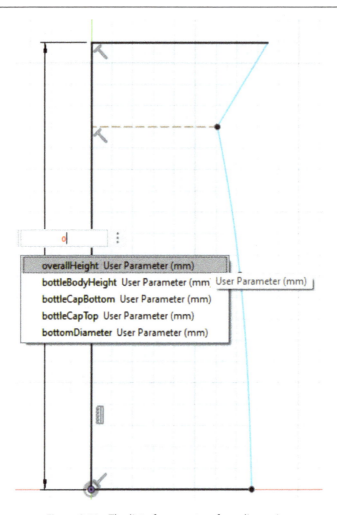

Figure 8.11 – The list of parameters for a dimension

4. Continue adding the user parameter dimensions as shown in *Figure 8.12*. Remember that you can place a dimension by left-clicking on a line edge, from one end point to another end point, or from an end point to an edge. Click **FINISH SKETCH** when you have finished adding the dimensions.

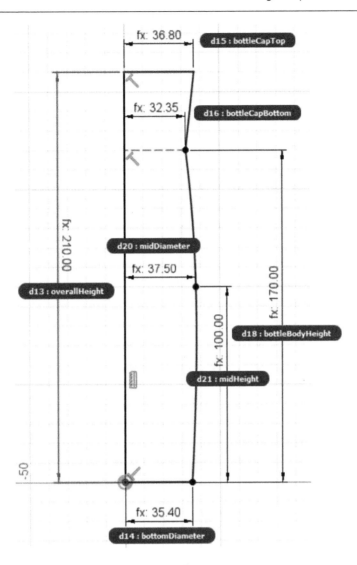

Figure 8.12 – User parameter dimensions added to sketch

Flexing the sketch

It's a good idea to test the parametric dimensions to make sure that they are working the way they are supposed to, which is called "flexing" the model. The way you flex a model is by going through each dimension and changing it to a different number to make sure that it acts the way that you believe it should work. Since we created two other bottles, let's change some of the dimensions to create another bottle size.

Let's change the parameters to those of the larger size bottle by doing the following:

1. Go over to the **BROWSER** tab on the left side of the screen and right-click on the sketch name.

2. Move your mouse down to **SHOW DIMENSIONS** and select the icon.

    This reveals the dimensions of a sketch without having to go back into the Sketch Environment.

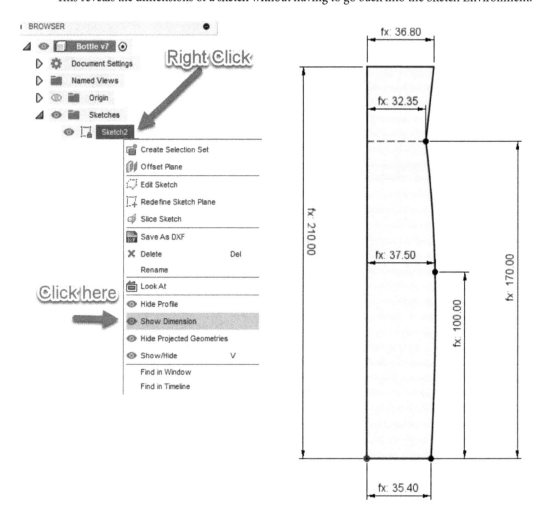

Figure 8.13 – The Show Dimension location

3. Go to the **MODIFY** drop-down arrow and select the **Change Parameters** command. This will open up the **PARAMETERS** floating window.

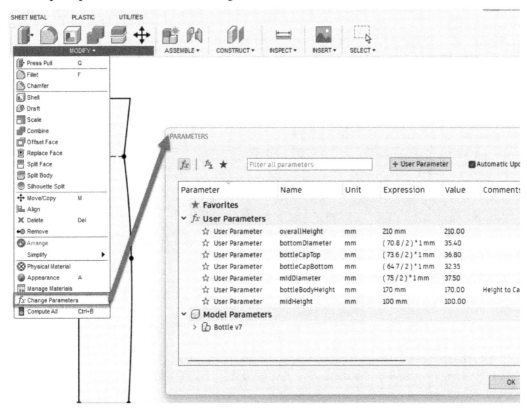

Figure 8.14 – The Change Parameters location

I like to change the overall height first since this will stretch out or shorten the model and will be less likely to break any dimensions. Let's first start by changing the **overallHeight** dimension.

4. Change **overallHeight** to **230mm** by left-clicking on the number in the **Expression** column.

5. Hit *Enter* on your keyboard to see the dimension change within the Fusion 360 sketch.

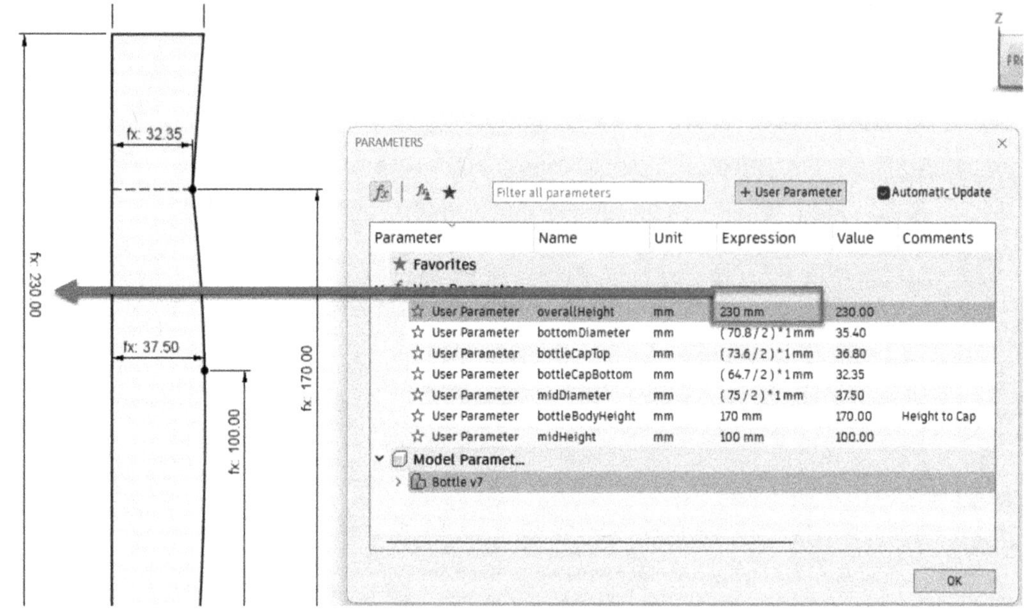

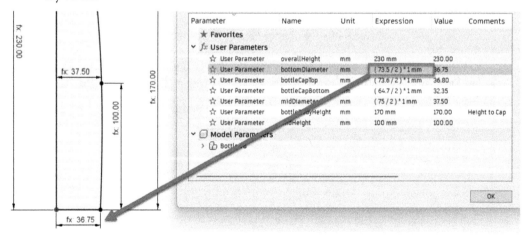

Figure 8.15 – Changing the overallHeight expression

6. Let's change **bottomDiameter** now. This expression has a formula within it because we needed to find the radius for our sketch, so we just need to change the first number of **70.8** to now say 73.5.

Figure 8.16 – Changing the bottomDiameter expression

7. For the next user parameter, we will create another parameter called `bottleCapTopFULLDiameter` and give it a diameter of `73.6`. This way, we can change a parameter and not accidentally change or remove any formulas:

a. Go to the **+ User Parameter** button within the **Parameters** flyout, add the name `bottleCapTopFULLDiameter`, and give it an expression of `73.6mm`.

b. Go to the **bottleCapTop** expression and only replace `73.6` within the `(76.2/2)` formula by selecting the relevant numbers and hitting *Delete*. Then, type in `bottleCapTopFULLDiameter`. A popup will appear showing the expressions, from which you can then select `bottleCapTopFULLDiameter`.

c. You may receive a warning to remove the mm at the end of the expression. If so, remove the mm at the end of the formula.

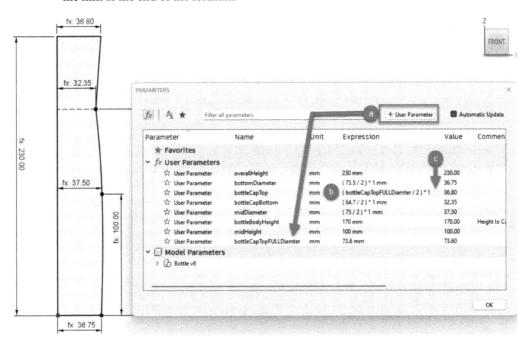

Figure 8.17 – Steps for adding a User Parameter into a formula

d. Now change `bottleCapTopFULLDiameter` from **73.6mm** to `43.9`mm. This will also change the `bottleCapTop` expression formula as well.

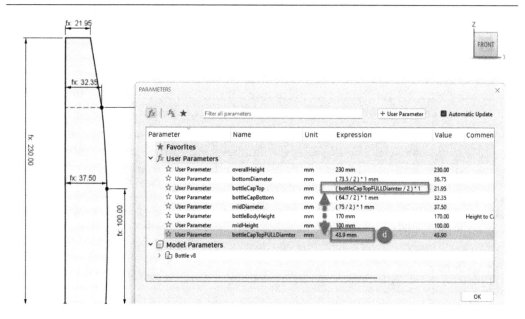

Figure 8.18 – Linking User Parameter expression values

8. Continue down the parameters list changing the values to the larger bottle size. I've added a screenshot so that you can match mine. I will leave it up to you whether you change the other parameter values to names or leave them as is. Click **OK** when you are finished to close out the **PARAMETERS** dialog.

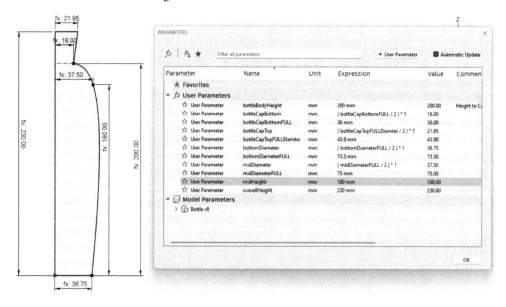

Figure 8.19 – Flexed bottle with added parameter names

Now that we have created our parametric sketch, we can start to create the 3D reference model by using the **Revolve** command.

Creating a 3D reference model

Another way to create a bottle 3D model would have been to use multiple sketches and then to use the Loft tool to generate a 3D model. Sometimes, that can complicate a model since we will have multiple sketches on multiple planes all at different heights, which can get confusing. Using one sketch profile can make a complicated model much easier to navigate: we will now use that single sketch to generate the 3D model using the following steps:

1. Select the **Revolve** tool within the **CREATE** panel. The profile should be automatically selected. For **Axis**, select the long vertical line that represents the center of the bottle.

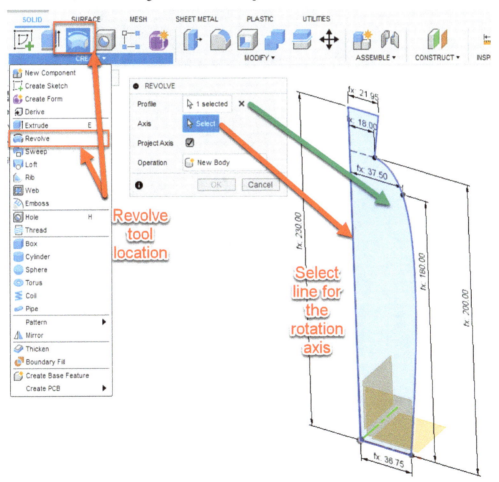

Figure 8.20 – Revolve tool location and axis selection

2.   After selecting the **Axis** line, another pop-up window will open with more options. We can leave these all at their defaults since we will need a full revolution of the profile to create the bottle. Click **OK** to close the **REVOLVE** dialog.

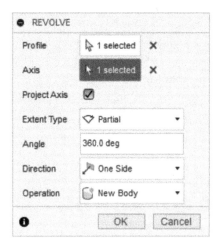

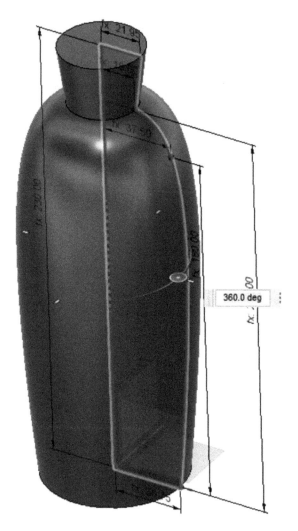

Figure 8.21 – Revolve tool options

Now what if we want to switch back to the midsized bottle that we created first? We will need to go back into the **PARAMETERS** dialog box to do that.

3. Open up the **PARAMETERS** dialog box and adjust the numbers to those in the screenshot in *Figure 8.22*.

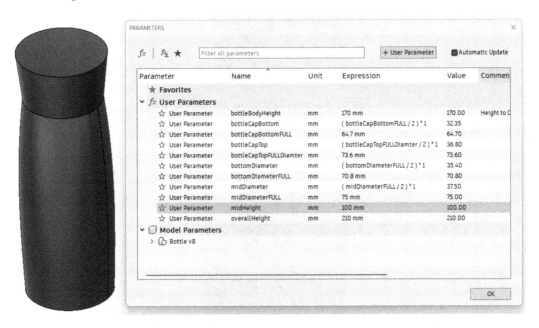

Figure 8.22 – Showing adjusted and flexed bottle

Perfect! The model flexes just the way we would like it to.

Now that you have finished the bottle, click on the **Save** icon in the top-left corner of your screen, and once you hit **Save**, you may notice that a small window opens asking for a **Version Description** value.

Figure 8.23 – Hit Save to add a Version Description entry

This is asking you to provide a brief description of what was accomplished in between saves. If you create a change where you want to remember what you did, hit **Save** and add a line of text describing what you did. This will help when you open the **Data** panel and want to go back to a specific previous version of your model.

Figure 8.24 – Data panel showing version history

Now that we have created both the bicycle handlebars and the water bottle models, we can put them together and reference them both in the same model.

Referencing the bicycle handlebars and water bottle models together

In Fusion 360, you can create components all located within the same model without having to pull externally referenced models, meaning I could have created the bicycle handlebars and bottle all in the same drawing if I wanted to. In this instance though, we created them separately and will bring them in as links into a new drawing:

1. If you still have the bottle design open, start a new Fusion 360 drawing by clicking on the *file* icon in the top-left corner and then **New Design**, or by clicking on the + tab icon in the top right.

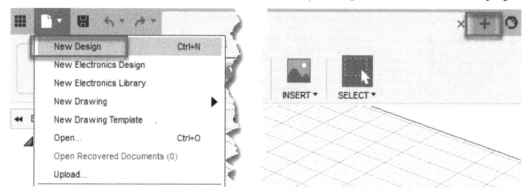

Figure 8.25 – New Design located within the file dropdown, and the location of the + icon

2. You will need to save this new file before you can insert other parts into it; otherwise, you will receive an error asking you to save the file first. Hit the **Save** icon and name the file `bicycle and bottle`.

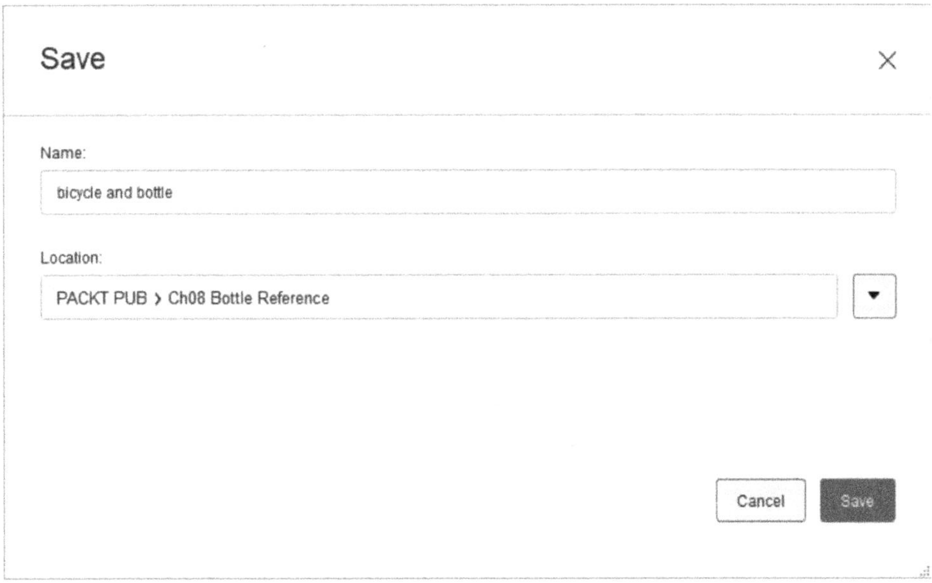

Save
Name:
bicycle and bottle
Location:
PACKT PUB › Ch08 Bottle Reference
Cancel
Save

Figure 8.26 – Save dialog window

3. Left-click on the **Data** panel to open it up and navigate to the previous chapter's `Ch07 Handlebars Reference` folder. You can get to the previous chapter's folder by clicking on the `PACKT PUB` name (red arrow).

Kevin Land
Data
People
Upload
New Folder
PACKT PUB
Ch08 Bottle Reference
bicycle and bottle
9:14:38 PM
V1
Bottle
1/25/23
V5

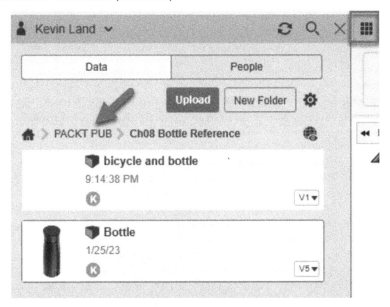

Figure 8.27 – Data panel open

4. Once you are within the Ch07 Handlebars Reference folder, right-click on the Bicycle Handlebars Fusion file and select **Insert into Current Design**.

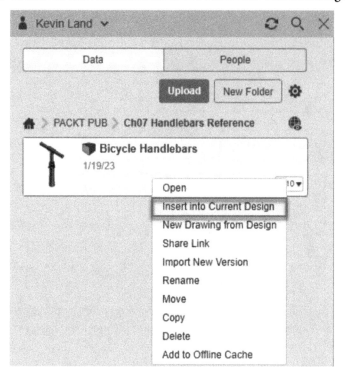

Figure 8.28 – Right-click on a file to get the Insert into Current Design option

5. A preview of the handlebars model will show up in your current bicycle and bottle file and the **MOVE/COPY** dialog window will appear. You can move, rotate, and scale this file using this flyout, but for now, just click **OK** to accept the default placement at *0,0*.

Figure 8.29 – The MOVE/COPY dialog window

6. Let's do the same for the bottle that we just created. Navigate to Ch08 Bottle Reference from the **Data** panel and right-click on the Fusion 360 Bottle file. Right-click and select **Insert into Current Design**.

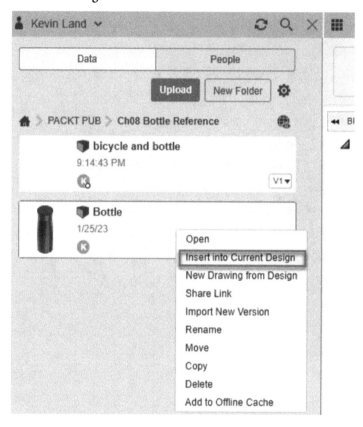

Figure 8.30 – Inserting the second component into the current design

7. The bottle model will be inserted into the same location as the handlebars. We will use the **MOVE/COPY** tool now to slide it over a bit, out of the way of the handlebars. First, left-click on the View Cube's **RIGHT** side (indicated by the red arrow in the following figure) so that we now see the model from the right. Now left-click on the horizontal drag arrow (indicated by the green arrow) and move the bottle to the left, or you can type in −2.8 into the **Y Distance** value box. Click **OK** to finish the command.

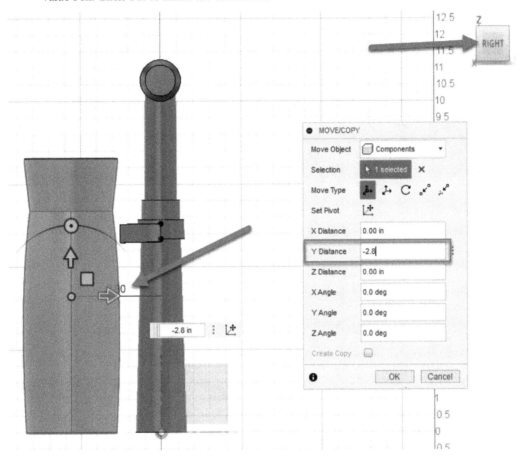

Figure 8.31 – Moving the bottle out of the way using the MOVE/COPY tool

Notice that there are now two components with chain icons within the **BROWSER** dropdown. This chain icon means that these files are now linked in the drawing file. This means that if the dimensions of the models were to be changed in those files, then they would be updated in this current file as well.

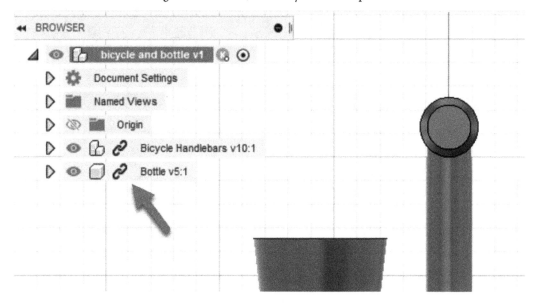

Figure 8.32 – The chainlink icon within the Browser

Hit the **Save** icon to finish the project and close the file.

Summary

In this chapter, we learned how to create a bottle for reference using a single sketch and the **Revolve** tool. We also learned how to insert previously created models into our current design so that they can be used as references. In the next chapter, we will create our water bottle holder, which will use the reference models that we designed in *Chapters 7* and *8*.

9

Building the Bottle Holder

We finally come to the last part of our multichapter build: creating the bottle holder. We will use both reference models from the previous two chapters and create a new model to hold our bottle in place on our bicycle.

In this chapter, we will learn how to bring in both reference models from the previous two chapters and then build a new model from those reference models. This will give us a good idea of how large to make the bottle holder and how it will attach to the bicycle. By the end of this chapter, you will have learned how to import a part into your current design, build a new part from an existing design, and add an as-built joint to show simple movement.

In this chapter, we're going to cover the following main topics:

- Designing and drawing out the model
- Creating the bottle holder body
- Creating the grip attachment

Technical requirements

You can practice with the files provided or create your own for a more custom experience. If you have difficulties with the project at any point, you have the ability to download the sample file and scroll through the timeline by dragging the arrow at the top of the timeline to see how it was created, which can make it easier to reproduce. The sample design for this chapter can be found at `https://github.com/PacktPublishing/Improving-CAD-Designs-with-Autodesk-Fusion-360/tree/main/Ch09`.

Designing and drawing out the model

The first thing to do when creating a new part is to draw what the new part may potentially look like. We can always change our minds later on, but it's always a good idea to get something down on paper about what you plan on creating. I did not include any dimensions since we will be using the existing bicycle handlebar and mid-size bottle to get the general shapes.

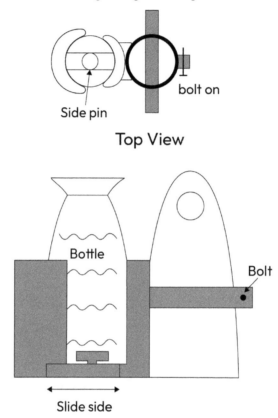

Figure 9.1 – Drawing of the bicycle part

Next, we need to figure out the design intent of this model. We want the model to be able to adjust to various heights of water bottles and various widths. I came up with an open-top piece that can adjust out with a slide bar at the bottom for variously sized bottles. If you too have a great idea for a bicycle water bottle support, now is the time to make it. Draw it out and try to create it in Fusion from what we've learned so far. The best way to learn is to try and fail and then try, try again.

Creating the bottle holder body

Let's get right into creating in Fusion 360. Start a new drawing, hit the save icon in the top left, and then create a new folder named Ch09 Bottle Holder. Save the file with the name Bottle Holder. Set the units for the drawing to millimeters instead of inches.

Figure 9.2 – The Save pop-out window

Now that we've saved, the first thing we need to do is bring in our references from the previous chapter:

1. Open the **Data** panel in the top-left corner and navigate to **Ch08 Bottle Reference**.
2. Right-click on the `bicycle and bottle` file and click on **Insert into Current Design**.
3. Click **OK** to accept the default location at the origin.

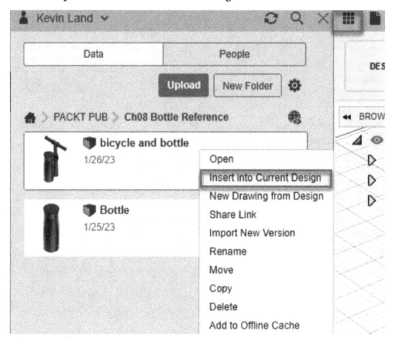

Figure 9.3 – Insert into Current Design

Since we are going to be working on multiple parts, we are going to need to create some components.

Creating multiple components

Remember Rule#1, which I mentioned in the first few chapters? The rule is to always create components for your multiple body parts. You may have noticed that I broke that rule a few times in the past few chapters when creating bodies, but it is most important when you are dealing with multiple parts that must come together to form a multibody part as you will need it for a **Bill of Materials** (**BOM**) or for creating joints, which we will do in this chapter.

Follow these steps to create multiple components for our bottle holder:

1. Click on the **ASSEMBLE** dropdown or the **Component** icon and create a new component.

2. In the **Name** field, type Holder 01 and click **OK**.

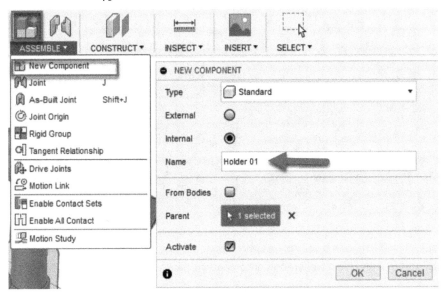

Figure 9.4 – NEW COMPONENT pop-out window

> **Important tip**
>
> Notice that your reference models have turned transparent. This is OK. This is what Fusion does when you have a new component created. It hides everything else, so that your new component body is easier to see.

3. In your browser, on the left side of the screen, mouse over **Bottle Holder v1** and left-click the active circle icon that appears on the right. This will set the top level to active. This will also turn the hidden reference bodies back to solid.

4. Add another component and name this one Holder 02. The reason why we need to go to the top level before adding in another component is to not create another component within Holder 01. We want to make it a separate component rather than a multi-component one.

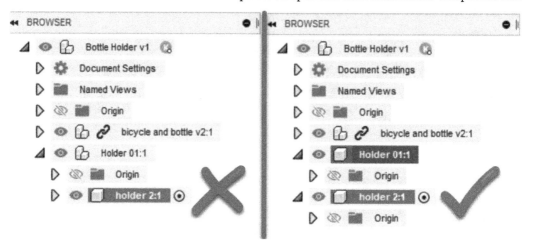

Figure 9.5 – The difference between placing a component in the top level, correct for this project, compared to placing it within another component, which is incorrect for this project

5. Activate Holder 01 by mousing over the name and when the active circle appears on the right, left-click on it to turn the circle solid. Your reference models will turn clear again.

Now that the components are set up, we can start creating the bodies within the components.

Creating the Holder 01 component

We will start with the support around the bottom of the bottle first:

1. Orbit your screen to see the bottom view under the bottle. Start the **CREATE SKETCH** tool and left-click the bottom circle of the bottle, not the plane.

   The reason why we are selecting the bottle instead of the plane is that if we decide to move the bottle up a few inches, the Holder 01 body will move with it. It won't move if the sketch is attached to the XY plane, which is why we are selecting the bottle face.

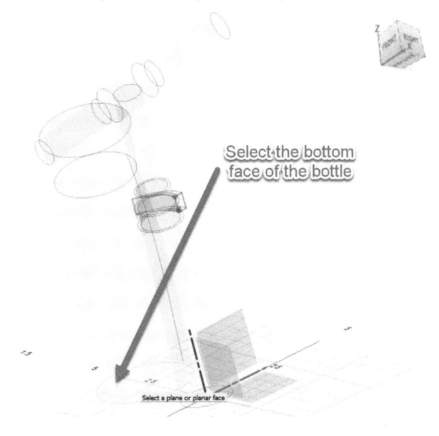

Figure 9.6 – Selecting the bottom face of the bottle

2. The screen has rotated to view the bottom face of the bottle, but it is hard to see what the bottom face is since we can see the rest of the bottle as well. Right-click and drag slightly to orbit and see the rest of the bottle. Now we can see where the bottom face actually is.

3. Click on the **OFFSET** tool located within the **MODIFY** panel and left-click the bottom circle edge. Set **Offset position** to 15 mm.

4. Click on **FINISH SKETCH**.

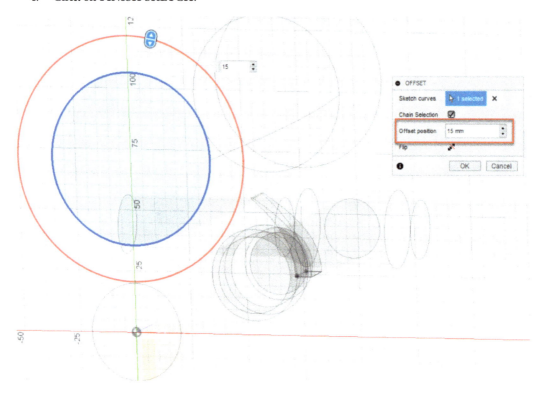

Figure 9.7 – Bottom of bottle and Offset set to 15mm

5. In the top-right corner of your screen, click on the **ViewCube Home** icon to set your view back to a perspective view.

6. Click on the **EXTRUDE** tool located within the **CREATE** panel and click on both sketch profile faces.

7. Extrude away 20 mm from the bottom of the bottle and then click **OK**.

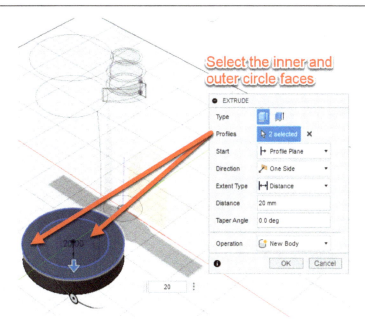

Figure 9.8 – Left-click on both sketch profiles and extrude 20 mm down

8.   Click on the **CREATE SKETCH** tool and select the YZ plane.

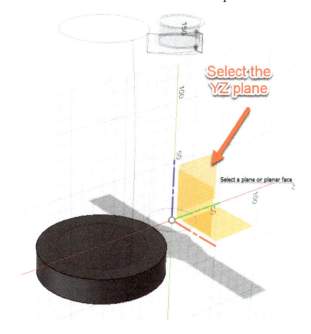

Figure 9.9 – Select the YZ plane to create the next sketch

9. Within the **CREATE** panel dropdown, select **Project / Include** and then select **Intersect**. Select both the water bottle and bicycle pole references. This will turn the lines purple, which means that they are now referenced in your current sketch so you can use that geometry to build upon.

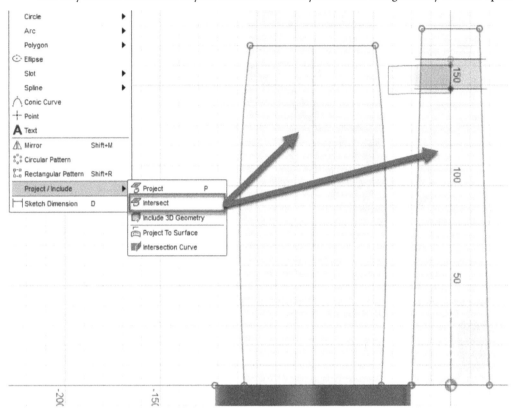

Figure 9.10 – The purple lines of the Intersect tool on the reference models

10. Within the **MODIFY** panel, select **Offset** and set it to 2   mm for the bottle edge.

    For the right edge, we will draw a line from the endpoint.

11. Select the **LINE** tool and try to pick the endpoint of the cylinder extruded base. You will notice that you may not be able to since there are no endpoints on a cylinder corner. So, instead, we will project in the top edge of the cylinder so that we can connect to the right-side endpoint.

12. Select the **LINE** tool once more, if you accidentally selected out of it, and pick on the edge of the cylinder (see *Figure 9.11*). This will automatically project in the edge, which will show a purple dot at the end that we can constrain to.

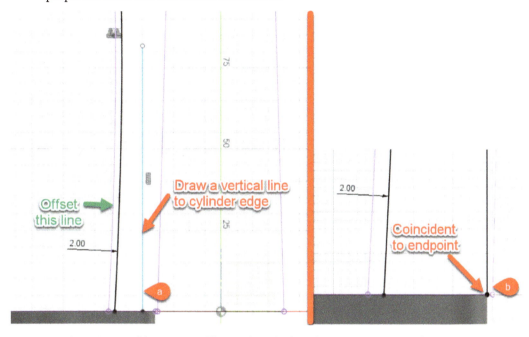

Figure 9.11 – Adding a vertical line to the cylinder edge to gain a projected endpoint

13. Now create a horizontal line from the top endpoint and set a dimension of 130 mm. Remember, if we don't like that height, we can always change it later. Be sure that you see a blue inner color as shown in *Figure 9.12*, otherwise you may have an open edge or an endpoint that may not be connected.

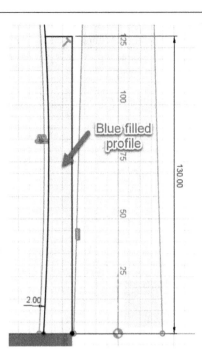

Figure 9.12 Adding a horizontal line with a dimension

14. Hit **FINISH SKETCH**.

    We will be creating a sweep for that last sketch, and to create a path, we need to have another sketch.

15. Create another sketch and select the top face of the extruded cylinder. Then create a projected circle of the top edge.

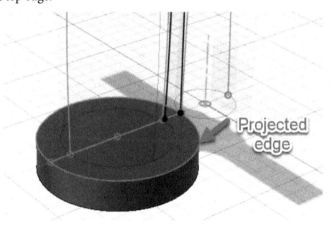

Figure 9.13 – Projected edge of top face

16. Select the **SWEEP** tool within the **CREATE** panel and select the vertical profile; then, for the path, select the circle edge.

17. Set both **Distance** fields to 0.15. Feel free to play around with these settings to see what they control. Notice that it is measuring a total distance of 1, which means that it will fully go all the way around the circle. Since we only need it to be around a little bit, we will set both distances to 0.15.

18. Click **OK** to finish the command.

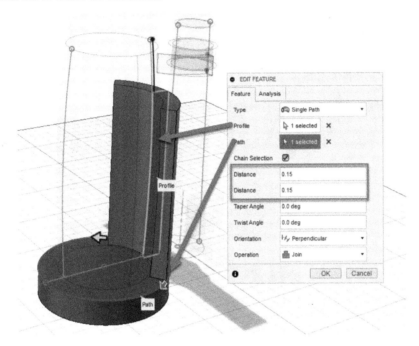

Figure 9.14 – The SWEEP tool and its selections

Now that we have the first body created, we will create the connection to the bicycle next.

Creating the grip attachment

Continue working within the Holder 01 container. If you exited Fusion and then open the project up once again, Fusion will start with the top level being active instead of the Holder 01 container. Keep a watchful eye on the browser and which container has the black dot.

Follow these steps to create the grip attachment:

1. Be sure that `Holder 01` is your active component.

2. Select the **OFFSET PLANE** tool and select the bottom face of the bottle holder. The reason why we are selecting that face is that if there are any changes made to the bottle holder body, such as its location, the grip will move as well.

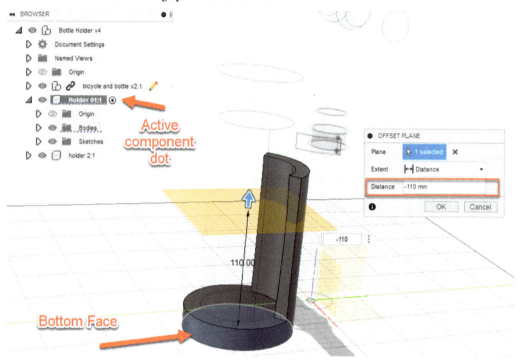

Figure 9.15 – The OFFSET PLANE tool with a distance of –110 mm

3. Create a new sketch on the new offset plane that we just created.

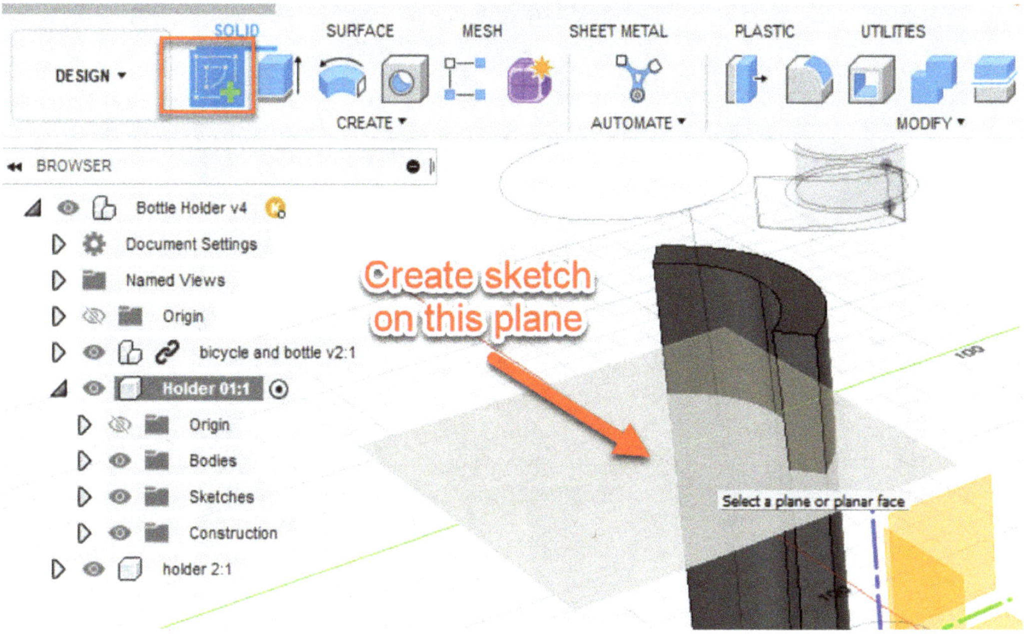

Figure 9.16 – Create the new sketch on the offset plane

The screen will now change to view the sketch plane from the top view. As we did previously, we need to pull in the edge of the reference bicycle pole.

4. Go to the **CREATE** panel drop-down arrow, select **Project / Include**, and select **Intersect**.

5. Orbit the 3D model a bit to see the pole better by holding down the right mouse button and dragging. Once you see the pole better, select it to see the purple reference circle.

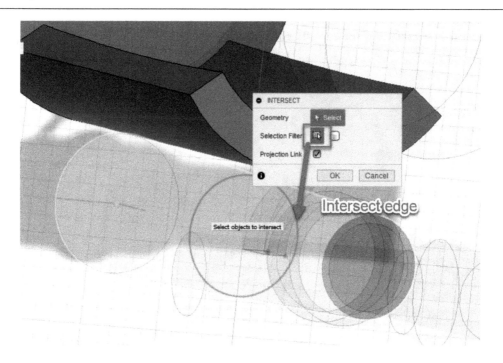

Figure 9.17 – The INTERSECT tool used on the bicycle pole

Let's set up a named parameter for this next step as we will need to offset that purple circle a small bit. We will need to add a tolerance to allow the grip to connect to the pole without issue.

6. Click on the **Change Parameters** icon located within the **MODIFY** panel.

7. Click on the **+User Parameter** button.

8. Enter the name as `tolerance`

9. Enter an **Expression** value of `1` mm.

    The reason why we are doing this is that if we 3D print this part and it is too small or too large, we can just go back to this expression and change the number and then reprint it quickly.

10. Click **OK** to close both windows.

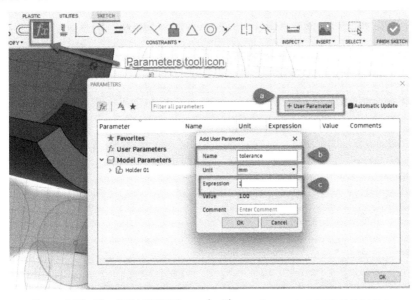

Figure 9.18 – The PARAMETERS panel with a custom user parameter name

11. Click on the **OFFSET** tool located within the **MODIFY** panel and then select the purple circle:

   a. For **Offset position**, type in the name `tolerance`.

   b. Make sure that the offset is growing away from the circle instead of going toward it. Click **OK** to set the offset.

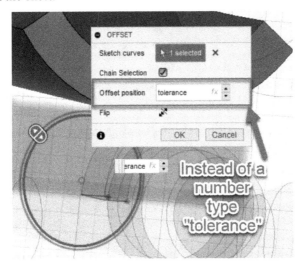

Figure 9.19 – The OFFSET tool with a parameter name

12. Now we will create the outer circle. Click once again on the **OFFSET** tool and try to select the offset that we just created.

Notice that we can't select the outer line, but we can still select the inner original circle. If you try to click on the circle, Fusion will throw an error stating that it is recommended to try another selection. This is designed this way because if the other offset was to be deleted, you would still have your other one untouched.

13. Click on the original line and set **Offset position** to tolerance+5. Doing it this way makes sure that if the tolerance number changes, we will still have 5 mm beyond that number. Click **OK** to finish the command.

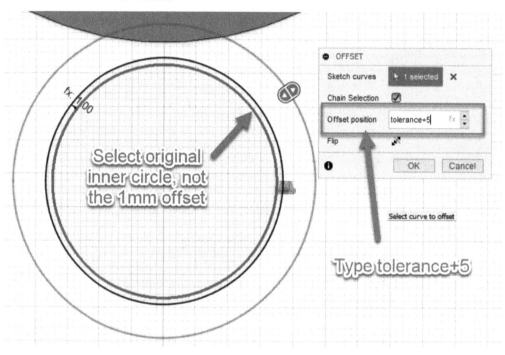

Figure 9.20 – The OFFSET tool distance with tolerance included

We will need to connect the exterior circle to the body, and instead of a straight line, we will use an arc.

14. Go to **CREATE** panel and then select the **Arc** flyout and then the **3-Point Arc** tool. When picking the three points, refer to *Figure 9.21*.

15. If the tangency constraints don't show up, be sure to add them by selecting the tangency constraint tool and picking on the circle and then the arc, not the endpoints.

16. Draw a horizontal construction line from the center dot to the right. We will connect to this line in the next step.

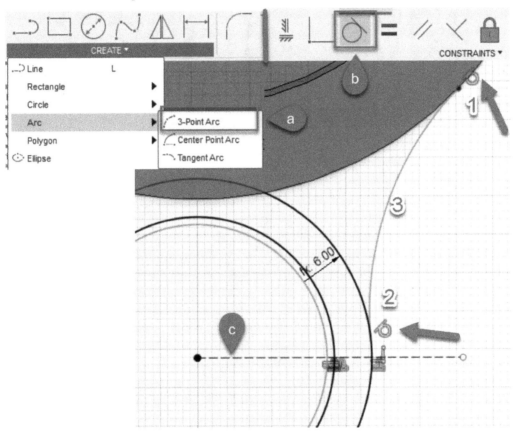

Figure 9.21 – The 3-Point Arc tool

17. Select the coincident constraint tool and pick the endpoint dot of the arc and the horizontal line (steps **1** and **2** in *Figure 9.22*).

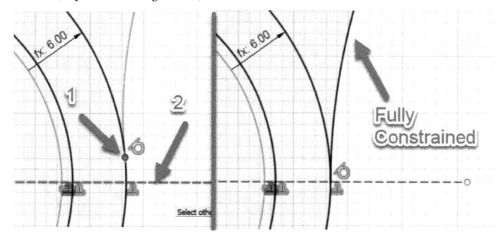

Figure 9.22 – Using the coincident constraint

18. Go to the **TRIM** tool located in the **MODIFY** panel and remove the excess line. You may get a warning saying that Fusion had to remove a constraint to get the **TRIM** tool to work correctly, which is OK.

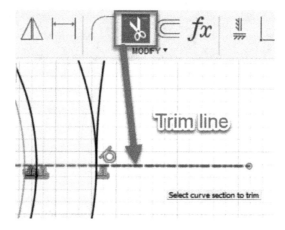

Figure 9.23 – The TRIM tool location

19. We will now mirror the arc to the left side using the following steps:

    a.   Draw a vertical construction line from the center dot.

    b.   Use the **MIRROR** tool to flip the arc to the other side. Click **OK** to finish with the tool.

    c.   Click **FINISH SKETCH** to return to the **DESIGN** environment.

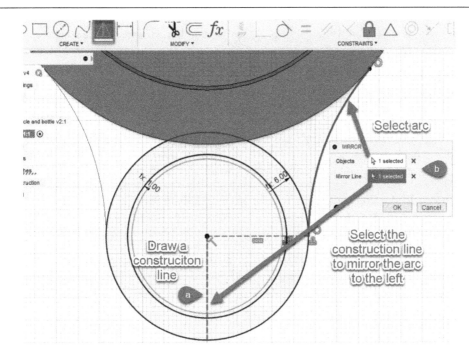

Figure 9.24 – The MIRROR tool's location

20. Click on the **Home** icon located on the ViewCube to orientate the screen to a perspective view.

21. Then, click on the **EXTRUDE** tool and select the three profiles located in *Figure 9.25*.

22. Set the extrude **Distance** value to 20 and be sure that **Operation** is set to **Join** since we want this grip to connect to the Holder 01 body. Click **OK** to finish the command.

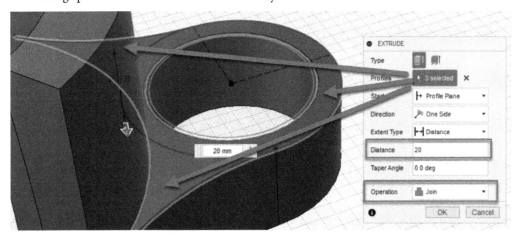

Figure 9.25 – EXTRUDE tool properties

Now that we've finished the bottle holder grip, we can start building the front slide attachment.

Creating the slider body

For the slider in the front, we will now make Holder 02 the active component and create another vertical bottle support. We will then move back to Holder 01 and cut out a section for Holder 02 to slide into as any geometric changes you make should be made within that component as it will make it easier to locate within the timeline. We will switch back to Holder 02 and create the slider portion and then add an as-built joint to it to allow it to slide. Follow these steps:

1. Make Holder 02 the active component by going to **BROWSER** on the left side of the screen, mousing over the Holder 02 name, and clicking on the dot that appears to make it solid. Your Holder 01 model will become transparent.

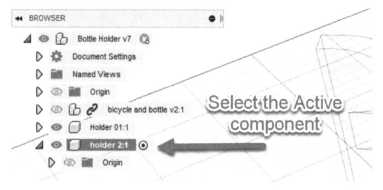

Figure 9.26 – Changing Holder 2 to the active component

2. Click on the **CREATE SKETCH** tool and select the top face of the bottle holder. Your screen will change to an overhead view of the bottle holder.

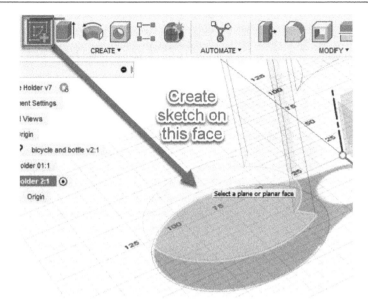

Figure 9.27 – Selecting the top face for the SKETCH tool

3. Draw a construction line using the **LINE** tool (use the keyboard shortcut of *X* to switch from a regular line) from the center straight down.

    We need to offset the bottom circle toward the center, but I'd like it to match the thickness from above. Instead of getting out of the command and going to look for that sketch, we can simply left-click on the edge and, in the bottom right-hand corner of your screen, you will see a distance measurement that says **12.997 mm**. You can do this to get information for any object by left-clicking on it and looking at the bottom-right corner of your screen.

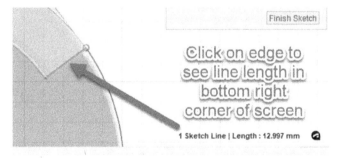

Figure 9.28 – Click on a 2D or 3D object to see its length

4. Click on the **OFFSET** tool to offset the outer edge 12.997 toward the center and be sure to unselect the chain selection box. This will make sure that only the arc and not all the connected edges will be offset.

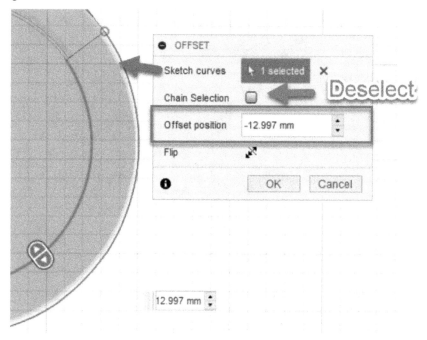

Figure 9.29 – Offset selection and deselecting the Chain Selection box

5. We will now add a post that will hold the bottle from falling:

   a. Draw a line on both sides that connects the outer circle and the inner circle.

   b. Use the symmetry constraint to make sure that both are the same distance by first clicking on the **SYMMETRY** tool, then clicking on **1** line, then the other line, and finally the construction line.

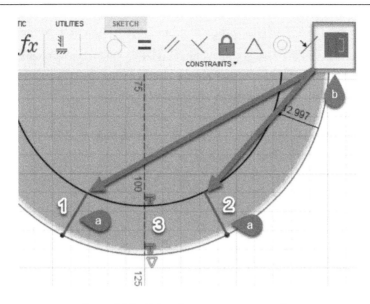

Figure 9.30 – Using the symmetry constraint

6. Now add in the dimensions with the keyboard shortcut *D* (see *Figure 9.31*):

    a. Select the line and construction line to create an angle dimension.

    b. Select the outer points of the same lines to create an angled dimension.

    c. Click on **FINISH SKETCH**.

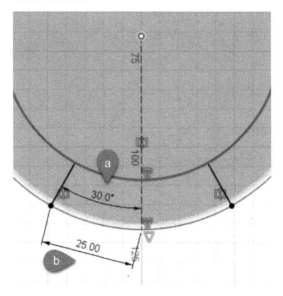

Figure 9.31 – Dimension locations for the sketch

We will now sweep this sketch using the path from the previous sweep on the opposite side.

7.   In the browser, turn on the eyeball icon within **Holder 01 | Sketches | Sketch 2** (your number may differ, but the sketch holds the profile of the bottle).

8.   Click on the **SWEEP** tool located within the **CREATE** panel and select the sketch and for the path select the curve.

9.   Set **Distance** to . 5 as we will have it be slightly smaller than the other side. Click **OK** to finish the command.

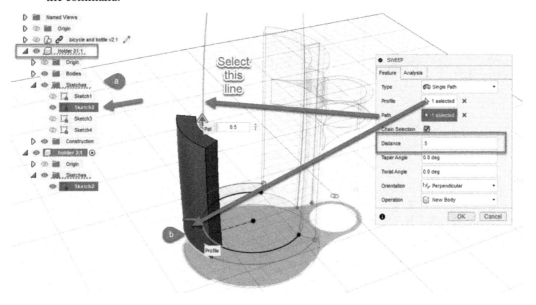

Figure 9.32 – SWEEP tool selections

Now that we have completed the extrusion support, we will cut an opening for Holder 02 to slide into.

Cutting a void space

You can create sketches to cut out material by changing the direction of an extrusion to cut into an existing 3D material; to do that, follow these steps:

1.   Go to **BROWSER** on the left and click on **Holder 01** to make it the active component.

2.   Click on the sketch and select the top face of the bottle holder. We will cut out a piece of this part, and since the cut action will happen to this body, it's best to make it the active component.

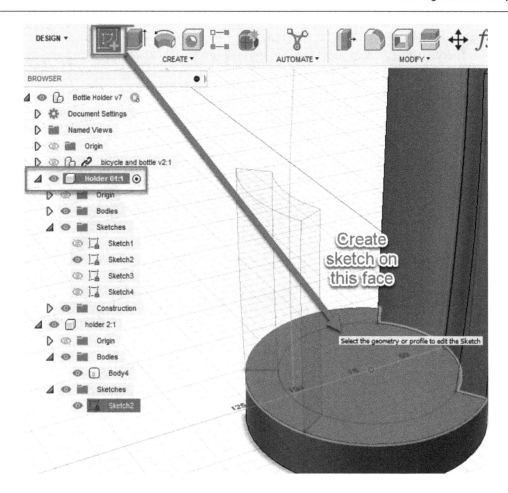

Figure 9.33 – Select the top face of the bottle holder

3. To create the cutout, follow these steps:

    a. Draw two vertical lines using the keyboard shortcut *L* from the top corner point of the smaller arc support that was just created straight down. The length doesn't matter but make sure that it has the vertical constraint attached if it wasn't automatically added (see *Figure 9.34*'s orange arrows). Now click on **Arc**, then **Tangent Arc**, and select both endpoints (red arrows).

    b. Drag the center point of this arc by left-clicking on it to match the center of the object (green arrow).

c.   Now finally click on **Project / Include**, select **Project**, and select the outer circle edge (purple arrow). When you select this, the inner area will be shaded in, which is a visual cue saying that this area is an active profile.

d.   Click on **FINISH SKETCH**.

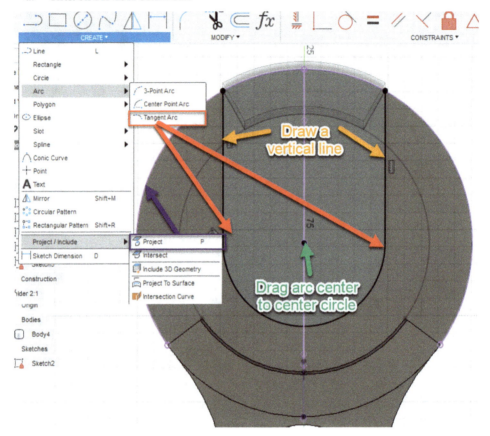

Figure 9.34 – Placing the sketch lines and arc and constraints

4.   Click on **Extrude** within the **CREATE** panel and select the profile that we just created. If you are having trouble selecting that profile, you can left-click and hold over the profile that you want, and a popup will display allowing you to select a specific profile. You also may have to select two profiles as the construction line down the middle may have split your profile in two. Set **Distance** to –10 mm and **Operation** to **Cut**, and then click **OK**.

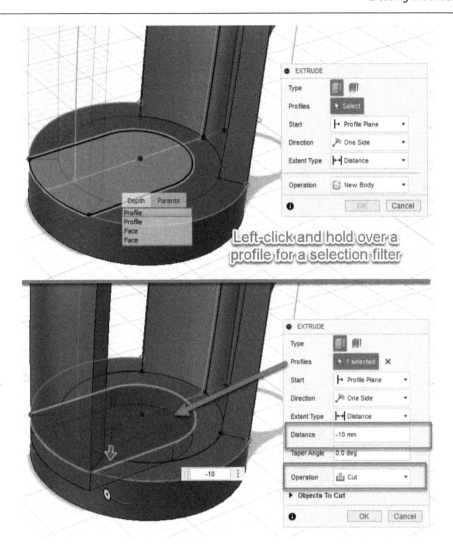

Figure 9.35 – Showing EXTRUDE options and selections

A void is created to allow us to create the mating part in `Holder 2` by using the same sketch.

Now we know how to create an opening in an existing 3D model. We will use the same sketch that created the void to create another separate body to fill the void by creating a mating part.

Using an existing sketch to create a body

We don't have to create another separate sketch to create a mating part in Fusion 360. We can use the same sketch that we used to create the opening to create another body. These are the steps to do that:

1. Go to **BROWSER** and left-click on the **holder 2:1** radio button to make that the active component.

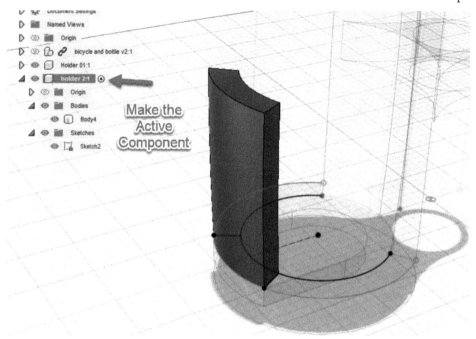

Figure 9.36 – Switching to the Holder 2 container

2. Open the browser and turn on the eyeball icon within **Holder 01 | Sketch5** (your number may differ). This is the sketch we used to create the void. Select the **EXTRUDE** tool with the **CREATE** panel and select the sketch. Set **Distance** to −10 mm and **Operation** to **Join**.

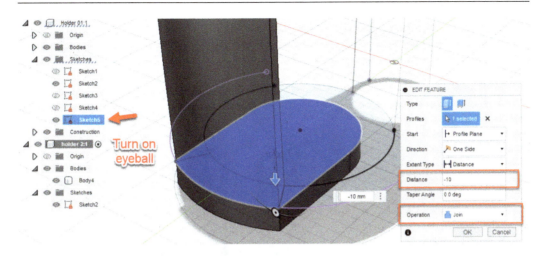

Figure 9.37 – Extruding using an existing sketch

We now have two separate bodies, one to grip the bicycle and the other for the bottle. Next, let's add an as-built joint to see the part move.

Adding in a joint

A great feature of Fusion 360 is adding joints to see parts move as they would in real life. One nice feature is what is called an as-built joint. What this means is that if you have two parts that are already placed where they would go and that are connected how they would be in real life, you can easily connect them with an as-built joint without having to worry about selecting multiple faces of different parts as you would in other design programs. Follow these steps to add an as-built joint:

1. Turn off all the sketches by going into **BROWSER** on the left side of the screen and selecting the top-level eyeball icon, which will turn off the remaining eyeball icons within that folder:

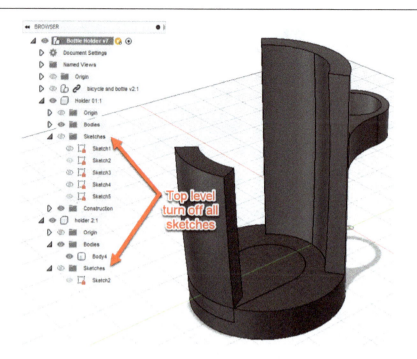

Figure 9.38 – Turn off the top-level sketches

2.   Go to the **ASSEMBLE** panel's drop-down arrow and select **As-Built Joint**.

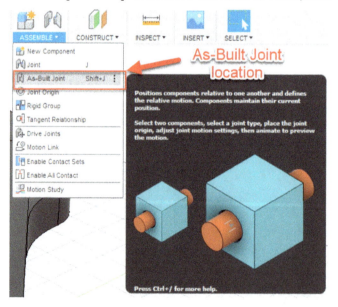

Figure 9.39 – As-Built Joint location and description

The **AS-BUILT JOINT** pop-up window will open.

3.   For **Components**, select both bodies by clicking on each part.

4.   For **Type**, click on the drop-down arrow to select **Slider**, since this part will slide back and forth.

5.   For **Snap**, select the circle at the top center face.

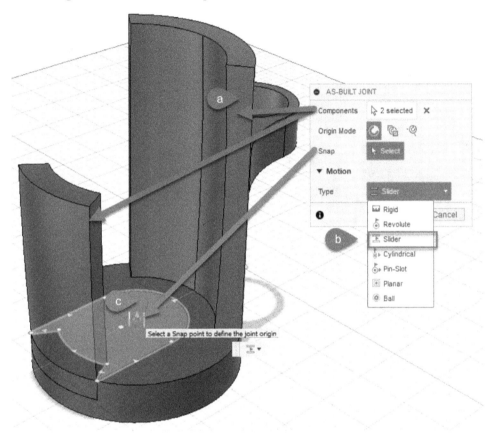

Figure 9.40 – AS-BUILT JOINT selections

6. After selecting the snap point, you will see a brief animation of what the slider will do. It defaulted to the *z* axis for my animation; yours may differ, but we want it to slide along the *y* axis instead. For the **Slide** option, select **Y Axis** and notice that the animation now changes and shows the part sliding in the correct way. We can set more options but for now, let's enjoy what we built and click **OK**.

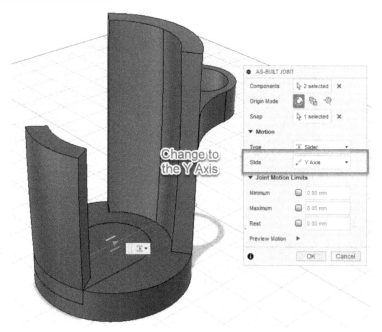

Figure 9.41 – Additional AS-BUILT JOINT options

7. If you try to click and drag on the part, you will notice that both parts move. We need to secure one down so that it doesn't move at all. Go to **BROWSER**, right-click on the `Holder 01` name, and select **Ground**. This will set that part in place so that it doesn't move. Now when you click and drag `Holder 2`, it will move!

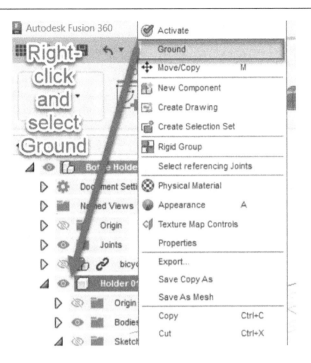

Figure 9.42 – The Ground tool's location

8. To bring the part back to its normal resting position, go to the top panel and select **Revert Position**. Be aware that this tool only shows up when a 3D model has moved to a new position. It asks to see whether you want to keep the current location or bring it back to normal.

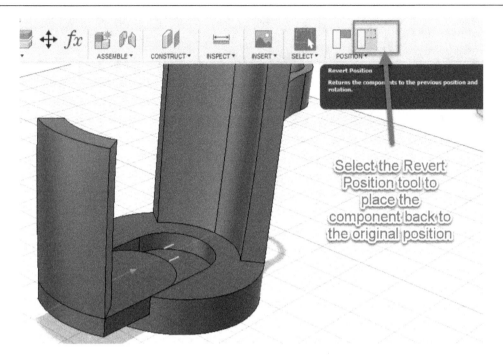

Figure 9.43 – The Revert Position tool's location

We have now completed our two parts, but we still need to make some adjustments to this model. We need to add in more tolerances and some external parts, such as bolts, to fix the part in place. We will do this in the next chapter and make some more improvements.

Summary

In this chapter, we learned how to create the bottle holder body, the grip attachment, and the slide plate for the front. Now that we've finished our bicycle bottle holder, what do you think of it? Can you make it better? What would you improve? There is always room for improvement in a design. It is an iterative process, which means that you may never get it on your first try. You may design something and then not like it, which is OK. You can always change it, or start fresh and design something new!

In the next chapter, we will explore how to improve upon this design by adding in external parts and how to change existing geometry.

10

Improving the Bottle Holder Design

Making changes, especially to a design that was started from scratch, will mostly present something that needs to be corrected. Sometimes, it may be things that we thought were cool at the time but now don't make sense, or maybe something needs to be added since we forgot to do it while it was being created. Luckily, **Fusion 360** allows us to not just go back in time within the timeline but also go back to previous versions and make them the most current so that we can try something different.

We are going to be working on the model from the previous chapter adjusting and modifying various parts using the timeline and other modification tools such as **Press Pull** and **Shell**. By the end of this chapter, you will know how to modify an existing part, add parts to your model from the **McMaster-Carr** online library, and create a simple 2D part drawing.

In this chapter, we're going to cover the following main topics:

- Making design changes
- Adding parts from McMaster-Carr
- Creating a part drawing

Technical requirements

You can practice with the files provided or feel free to create your own for a more custom experience. The sample design for this chapter can be found at `https://github.com/PacktPublishing/ Improving-CAD-Designs-with-Autodesk-Fusion-360/tree/main/Ch10`.

Making design changes

We will first start with some simple design changes to our model such as adding in tolerances so that the slider will easily go in and out and also removing some added material to make the bottle holder a little bit lighter.

Adding tolerances

For those who may not be aware, a tolerance is a small gap that allows a part to slide or move without scraping against another part. The amount of tolerance depends on what you are trying to accomplish. For instance, we want the `Holder 2` part to have a small tolerance so that it doesn't get stuck when sliding it in or out. Let's start with changing that first, as follows:

1. Open the **Ch10 Holder Improvements** folder and open the file that is in that folder.

    This is a continuation from *Chapter 9* but notice that since it was copied from *Chapter 9*, the version starts at `v1`. So, if you start a new design in a different folder, it will start off like it was new and not keep its design history. This is OK for us as we will be making changes to this new version and moving to another version later.

    The following screenshot illustrates the process:

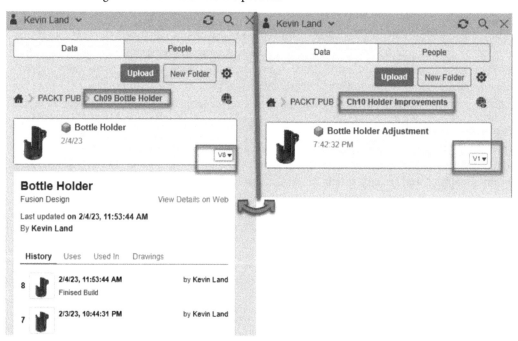

Figure 10.1 – Version history in a new design file

2. Turn off the eyeball icon within the **BROWSER** section for the bicycle and bottle container and for the Holder 2 container:

Figure 10.2 – Turning off two containers

We will use a direct modeling technique that will allow us to remove or add material directly to a model without having to locate the sketch that created it.

3.   Click on the **Press Pull** tool within the **MODIFY** panel and then select the faces where Holder 2 would slide in. Type in -1 in the **Distance** area to add a 1 mm gap. Click **OK** to finish the command:

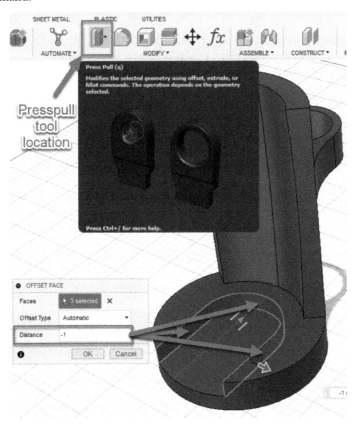

Figure 10.3 – Using the Press Pull tool to add extra spacing

4. Turn back on the eyeball for Holder 2 and notice that a small gap has been added around the area that we quickly changed:

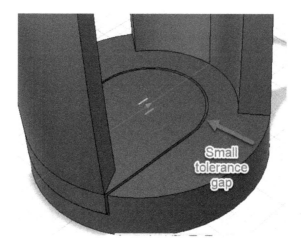

Figure 10.4 – Showing the added gap with Holder 2 turned on

> **Important note**
>
> The **Press Pull** tool allows you to change many different types of faces. You can change a corner edge to quickly have a smooth fillet by selecting the edge, remove extra material by selecting and dragging it inward toward another face, or add a hole by dragging all the way through.

Now that we can add a small tolerance gap so that the parts fit well together, we are going to remove some extra material from the part to not only make it less heavy but also to make it easier to 3D-print.

Removing material with the Shell tool

Too much material can weigh down a part, and if this part is being 3D-printed or made in a fabrication shop, it could cost more to create. A quick way to remove extra material is to use the **Shell** tool. Let's see how to use this tool to remove the extra material from our bottle holder:

1. Make Holder 01 the active component since we will remove extra material from this body:

Figure 10.5 – Holder 01 is the active component

2. Orbit to view the bottom of the Holder 01 part by either holding down, right-clicking, and dragging or using the ViewCube and clicking on the bottom-right corner:

Figure 10.6 – Orbiting to view the bottom of Holder 01

3.   Select the **Shell** tool by going into the **MODIFY** panel and then select the bottom face of Holder 01. Set the **Inside Thickness** value to 2 mm and notice that you get a preview of the **Shell** tool. Click **OK** to finish the command:

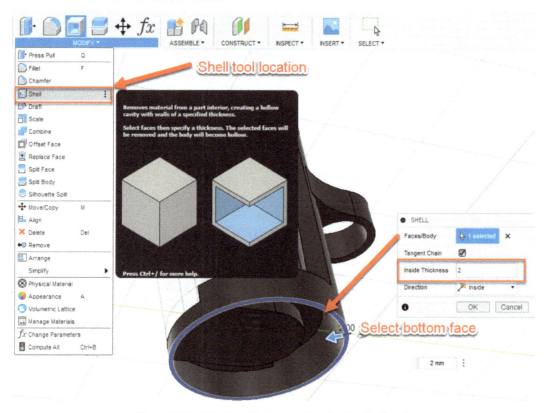

Figure 10.7 – Selecting the bottom face for the Shell tool

We now have a much lighter part that will print much quicker and use less material. We now need to add some parts from McMaster-Carr, which is an online parts store that has many 3D parts already created in its library for us to use for our designs.

Adding parts from McMaster-Carr

We will need to add a slider for Holder 2 and a screw for the rear mount to connect to the bicycle pole. We will start with the slider first. We need to first add a cut in the top of Holder 2 so that the screw will be set in a small amount and not interfere with any bottle that we place on top of it. Let's get started with the steps:

1. Remove the **As-Built Joint** slider that we added in the previous chapter by right-clicking on the **Slider1** name and then selecting **Delete**. We will add it again once we add in the McMaster-Carr parts:

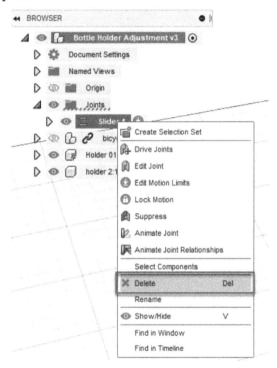

Figure 10.8 – Removing the As-Built Joint slider

2.   Activate Holder 2 and then create a sketch on the top face of the bottle holder:

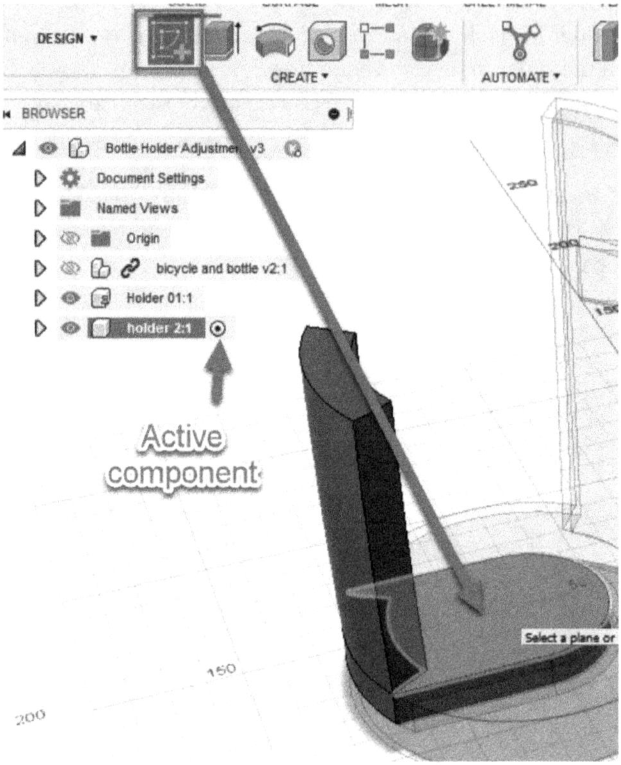

Figure 10.9 – Holder 2 activated and sketch face selection

3.   Go to **CREATE | Polygon | Circumscribed Polygon**. Select from the center point dot (red arrow) and drag horizontally, and type in 8mm to set the size.

We will need to add in a vertical constraint since we don't want this polygon to rotate.

4. Create a construction line and add it in from the center to an endpoint, then add in a vertical constraint if one is not added automatically, which will turn all lines black. Click on **FINISH SKETCH**:

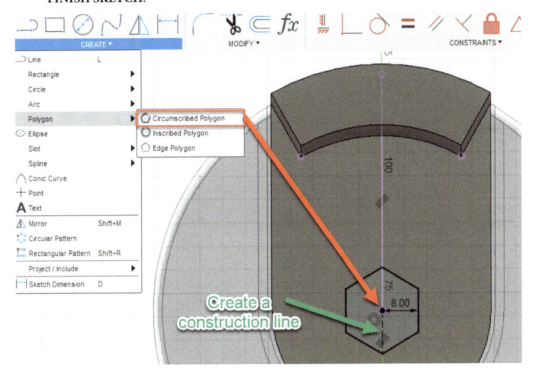

Figure 10.10 – Adding in a circumscribed polygon

Important note

The **Polygon** tool has three options within the drop-down arrow. The difference between them all is that a circumscribed polygon is created from the center to a midpoint edge.

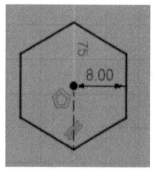

An inscribed polygon is created from the center to a corner point and will be slightly smaller than the circumscribed polygon.

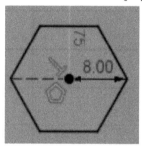

An edge polygon is created from corner point to corner point and all edges will be exactly the same length.

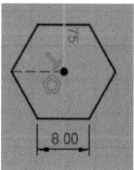

5.   Click on **EXTRUDE**, select the sketch that was just created, and set the **Distance** value to −5mm and the **Operation** type to **Cut**:

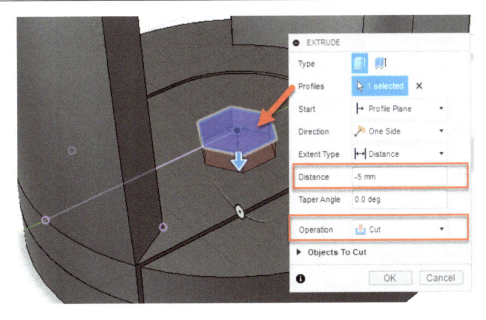

Figure 10.11 – Extrude to cut -5 mm

6. Create a sketch on the bottom face of the extrude that was just created:

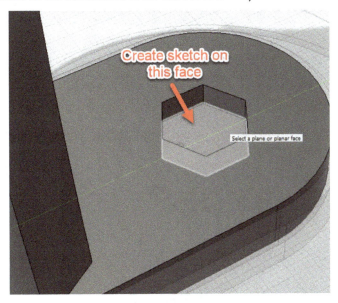

Figure 10.12 – Sketch face selection

7. Turn on the previous sketch so that you can reference the center dot—we want to be able to connect it to the polygon so that if the polygon should happen to move to another location, the circle will move too. Create a circle with an 8 mm diameter and click **FINISH SKETCH**:

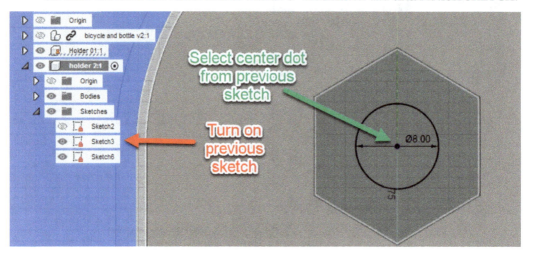

Figure 10.13 – Adding a circle sketch

8. Select the **Extrude** tool and then select the circle that was just created as our profile (if you have any difficulty selecting the circle profile, turn off the sketch that we previously used to reference the center dot).

9. Set the **Extent Type** field to **All** so that the cut goes all the way through both parts.

The other reason why we are choosing **All** is because if the bottom size should change, that cut through all will always cut through all pieces. This works much better than adding a numerical distance since that will only stay at a certain distance. Click **OK** to finish the command:

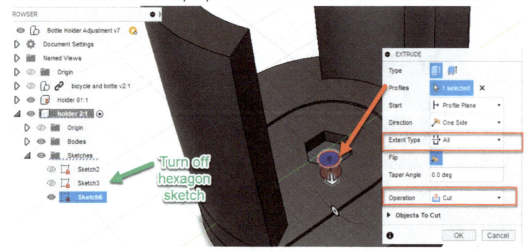

Figure 10.14 – Adding an extruded cut through all

10. We will add a slot to the bottom of Holder 01 so that our Holder 2 holder can slide. Orbit the model to see the bottom view and make Holder 01 the active component:

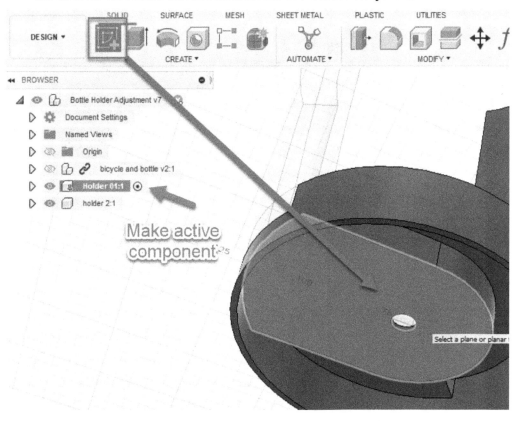

Figure 10.15 – Holder 02 active component and creating a sketch on the bottom face

11. Go to **CREATE | Slot | Center to Center Slot** and pick the circle that we created previously and then pick another point vertically below it, then left-click, and drag out a random diameter size for now:

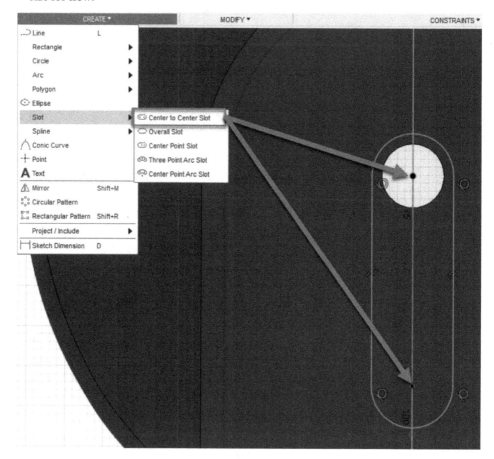

Figure 10.16 – Placing the center to center slot

12. Select the **Tangent** constraint and select the slot and then the circle (the red arrow in *Figure 10.17*). This will connect the slot diameter to the circle above so that if that circle diameter changes at any point, the slot diameter will change as well.

13. Now, add in a dimension with the *D* shortcut key and select the top and bottom circle points, and set a distance of 3 0mm.

14. Select the purple reference line and set it to be a construction line by selecting it and hitting the *X* keyboard shortcut (green arrow). The reason why we are setting this to a construction line is that it will be selectable as a normal line and will split up our slot profile. Setting it to a construction line will save us from having to select two profiles to extrude. Click on **FINISH SKETCH**:

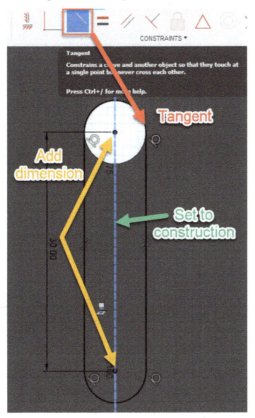

Figure 10.17 – Adding a Tangent constraint to the circle and a 30 mm dimension

15. Orbit your model to view the top of Holder 01 and turn off Holder 2 so that we can see the top face. Select **EXTRUDE** and pick the slot sketch we just created, set the **Extent Type** field to **To Object**, and select the top face. The extrusion will now only cut through until it reaches this top face. Click **OK** to finish the command:

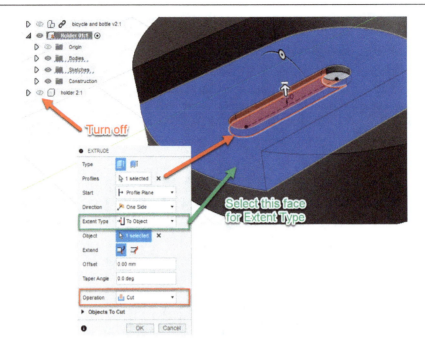

Figure 10.18 – Turning off Holder 2 and setting a cut extrusion

Now, let's add in a nut and a bolt from McMaster-Carr.

Adding a nut and bolt from McMaster-Carr

First, make the top-level component active by clicking on the bottle holder's name. The reason why we are activating the top level is that this part will be inserted as its own component. If you have Holder 2 active, it will be placed within that part, which is not what we would like to do:

Figure 10.19 – Making the top-level component active first

Follow these steps to insert a part from the McMaster-Carr online part library:

1. Orbit the model to the bottom view, go up to the **INSERT** drop-down arrow, and choose **Insert McMaster-Carr Component**:

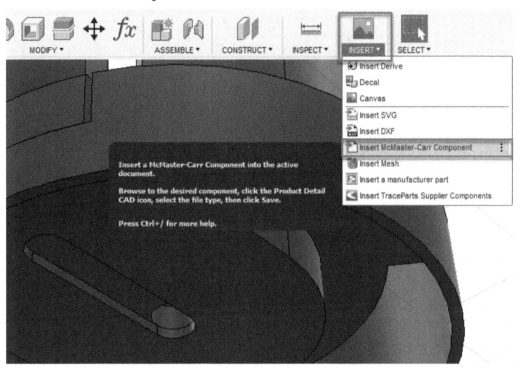

Figure 10.20 – The McMaster-Carr Component location

A McMaster-Carr pop-up window will appear and open a default page:

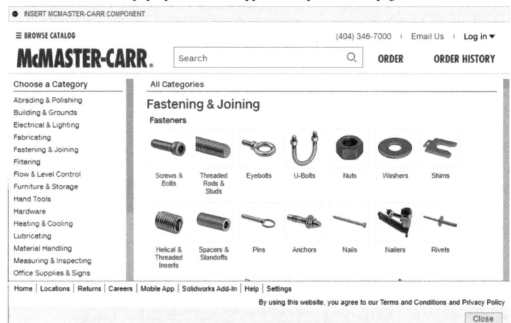

Figure 10.21 – The McMaster-Carr default page

2.   Click on **Screws & Bolts**, then scroll down and click on **Thumb Screws**, then click on **Metric Stainless Steel Low-Profile Knurled-Head Thumb Screws**:

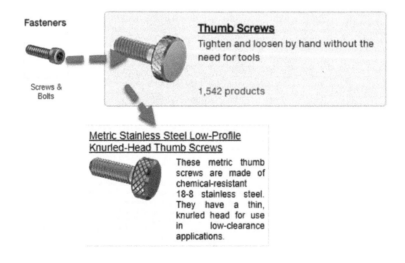

Figure 10.22 – Navigating to the thumb screws in the McMaster-Carr popup

3. Use the filter on the left to select an **8mm** head diameter value and an **11mm** length value. Now, choose the **M3 x 0.5 mm** thumb screw by clicking on part number **92545A119**:

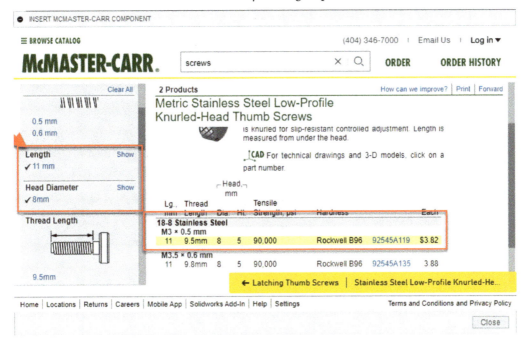

Figure 10.23 – Using the filter on the left to pick the right part

4. Another selection will appear asking which 3D model you would like to use. Select the drop-down arrow and pick **3-D STEP** and then click on **Download**:

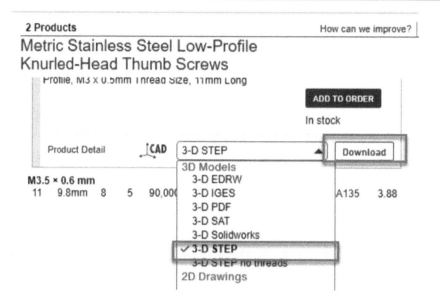

Figure 10.24 – Downloading the 3D McMaster-Carr part

5. The McMaster-Carr window will disappear, and the part that you chose will appear in your Fusion 360 space with a **MOVE/COPY** pop-up window. Click **OK** on this window as we will use joints to place this part:

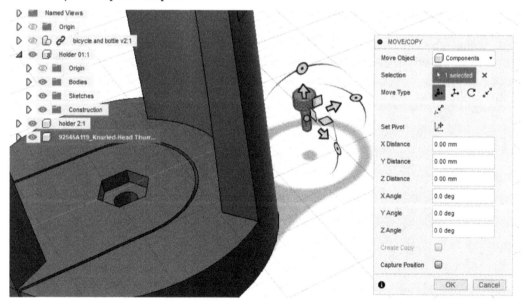

Figure 10.25 – The 3D McMaster-Carr part has been inserted

Next, we will use a joint to place this part on the bottom face within the circle.

Using joints to place a McMaster-Carr part

Now that we have brought in the part, we will need to place it in the slider hole. To do this, we will need to use a joint that will lock the part in place so that it will move accordingly. There are a variety of joints within Fusion 360, but the one we will be using is called a slider joint. Follow these steps to place the part:

1. Orbit your model to see the bottom of Holder 01 and click on **ASSEMBLE | Joint**:

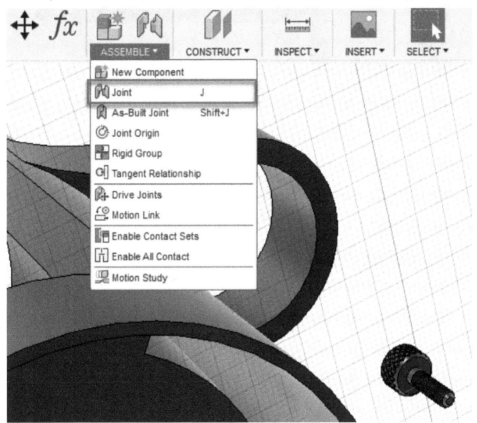

Figure 10.26 – The Joint tool location

2. A **JOINT** pop-up window will appear. Select the **Motion** tab and then click on the dropdown and select **Slider**:

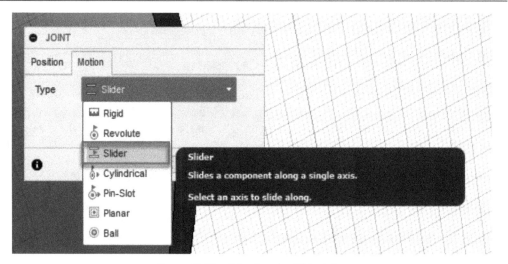

Figure 10.27 – Selecting the Slider joint

3. Select the **Position** tab and go under **Component 1** for the **Snap** selection. Click on **Select** and pick the bottom face of the screw:

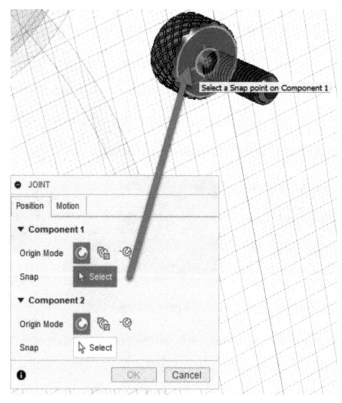

Figure 10.28 – Component 1 selection

4. For **Component 2**, pick the circle center point. Be sure to pick on the Holder 2 circle and not the slot circle as this will connect it to the bottom of Holder 2 and not the top part. If it is hard to pick, turn off Holder 2 to give yourself a better selection:

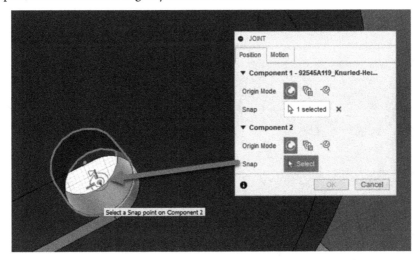

Figure 10.29 – Component 2 selection

5. The part will now move into place, but it may or may not appear in the correct orientation. If it did not, as mine did, click on the **Flip** button to correct its placement:

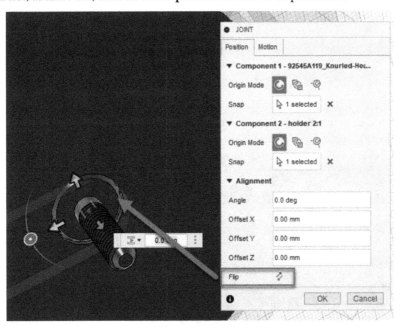

Figure 10.30 – Flipping the orientation

6. Adjust the **Offset Z** height to **2** mm to move it down and away from the bottom of Holder 01 so that it clears that model (you may have to place a negative sign in front of the 2—that is, -2—in order to adjust the screw). Something you will notice in most modeling software is that it will not tell you that you are hitting something while you are creating. There is a tool, located under the **INSPECT** drop-down arrow, named **Interference** that will show you if there is a collision, but you will have to activate it:

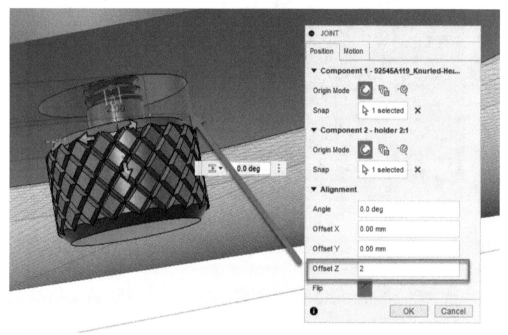

Figure 10.31 – Adding a Z offset to move the part from colliding

7. Finally, go back to the **Motion** tab and set the **Slide** field to **Y Axis** and check the **Rest** position of **0 mm** and a **Maximum** value of **30 mm** under **Joint Motion Limits**. Click **OK** to finish:

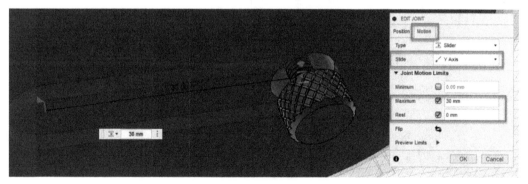

Figure 10.32 – Setting the slide and joint limits

If you click and drag on the part, it will move along with Holder 2, but it will move all over the place. We need to adjust the hole size and enable contact sets so that they will stay in place. Let's go back into the sketch and fix that, which will also fix our slot as well.

Fixing the hole size

The hole was created a bit too large so to adjust the size, we will need to go back into the timeline and shrink it down a bit. Follow these steps to accomplish this:

1. Go to the **BROWSER** section on the left side of the screen and navigate down to Holder 2 and right-click on the sketch that contains the hole (mine is **Sketch6** but yours may be different), and select **Edit Sketch**:

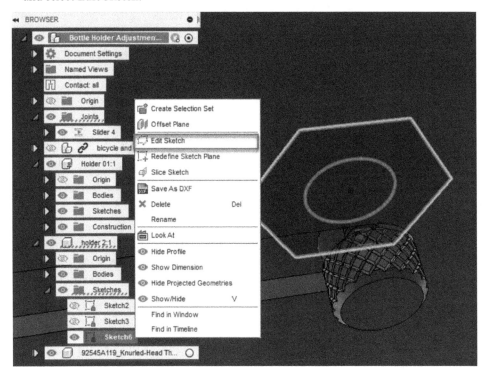

Figure 10.33 – Editing the hole sketch

2. Double-click on the diameter text and set the diameter to 4mm. Click on **FINISH SKETCH**. The slot and hole sizes have now changed to a smaller size:

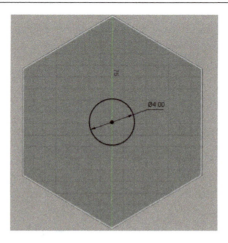

Figure 10.34 – Changing the diameter in a previous sketch

We will do the same for the hex nut size as well.

Bringing in a hex nut from McMaster-Carr

Let's locate a hex nut part in the McMaster-Carr catalog, set it with a joint, then adjust the hole to fit the nut. Follow these steps:

1. If we go to the **INSERT MCMASTER-CARR COMPONENT** tab and type in the number from our previous part, you will see that the screw thread size is **M3 x 0.5mm**. We will need that size for our hex nut:

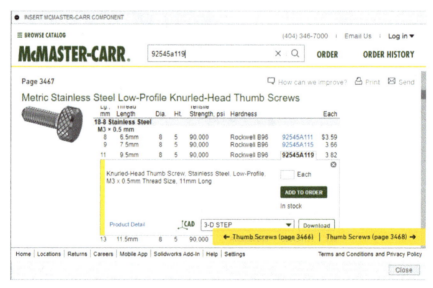

Figure 10.35 – Finding the hex nut size in McMaster-Carr

2. Click on **Nuts** and then use the left-side panel to filter down to the part that we want, as follows:

    - **System of Measurement** is **Metric**

    - **Thread Size** is **M3**

    - **Thread Pitch** is **0.5 mm**

    - **Thread Type** is **Metric**

    - **Nut Type** is **Hex**

    - **Hex Nut Profile** is **Standard**

    - **Height** is **4mm**

    - **Width** is **5.5mm**

    Filtering this way will lead us to part **90576A102**. Click on **3-D STEP** and then click on **Download** and, as before, click **OK** since we will place the hex nut part using joints:

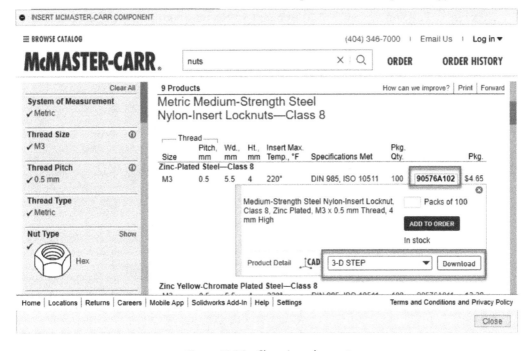

Figure 10.36 – Choosing a hex nut

3. Click on the **ASSEMBLE** drop-down arrow and then choose **JOINT**. Click on the **Motion** tab and set the **Type** field to **Rigid**:

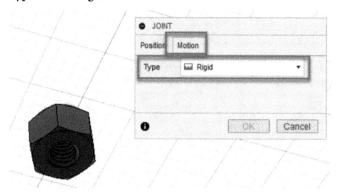

Figure 10.37 – Selecting the Rigid joint

4. Click on the **Position** tab, and for **Component 1**, select the bottom circle of the hex nut. If you're having a hard time picking the center, circle hold down the *Ctrl* key and move your mouse close to the center point, and it will snap to the center or any other points that it believes could be a snap location:

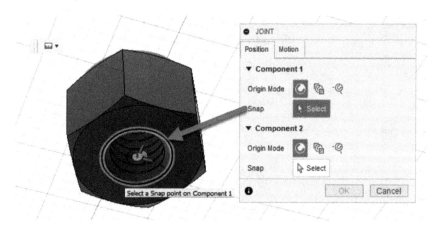

Figure 10.38 – Snap selection for Component 1

5. Orbit your model to the top face and select the center circle. Click **OK** to finish the command:

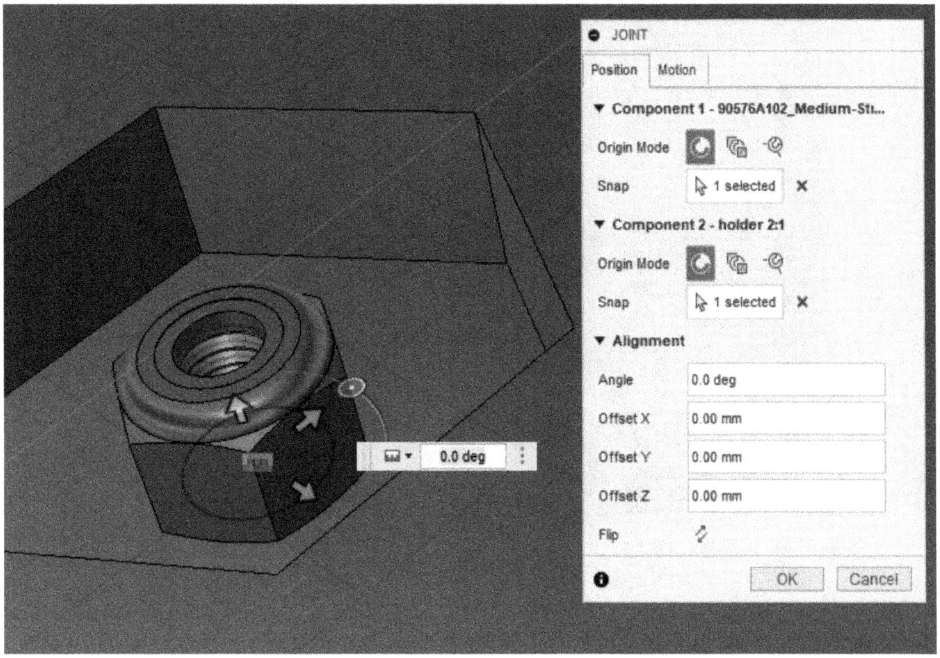

Figure 10.39 – Selecting the second snap point location

6. Now, we can adjust the size of the hex hole to match the hex nut size. Go to the **BROWSER** section, right-click on Holder 2 | **Sketch3**, and select **Edit Sketch** (your number may be different):

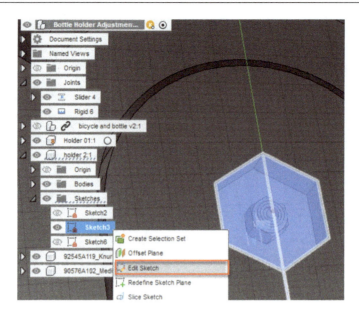

Figure 10.40 – Selecting the hex hole sketch

7. Set the new radius to 3mm, which will be 6mm in diameter—just enough to have enough space around the 5.5 mm hex nut. The reason why we have this hex nut hole is that when we adjust the screw on the bottom, the hex nut will stay in place while we adjust the screw. Click on **FINISH SKETCH**:

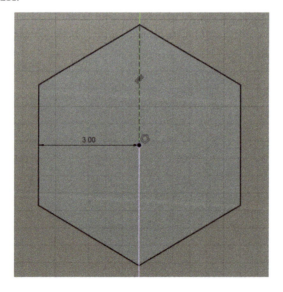

Figure 10.41 – Editing the size of the polygon to be 3 mm instead of 4 mm

Now, we can select Holder 2 and move the part along with the screw! Notice though that although the part moves around, it moves away from Holder 01:

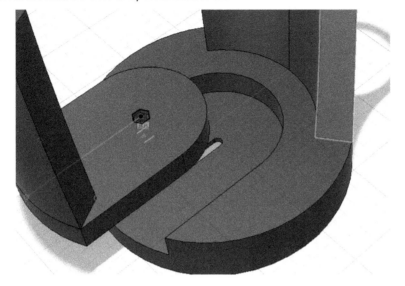

Figure 10.42 – Showing the joints are connected but the parts move away from each other

We can fix that by using another slider joint to connect Holder 01 to Holder 2.

Adding a slider joint for Holder 01 and Holder 2

Follow these steps to add a slider joint:

1. Click on the **Revert position** button or click **Undo** to get your model back to its original location. This option appears when a part has been manually moved around. You have the option to either keep the position as is or revert it back to the original location:

Figure 10.43 – Showing the Revert position location

2. We need to fix the slider that we placed on the bolt. That joint should be a rigid joint instead, and the slider joint needs to be placed on the holders. To fix a joint, right-click on **Slider** and select **Edit Joint** within the **Joints** folder located in the **BROWSER** section:

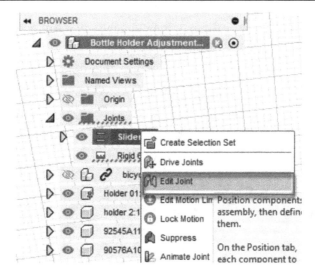

Figure 10.44 – Showing the location of the Edit Joint tool

3. An **EDIT JOINT** pop-up window appears. Select the **Motion** tab and change the **Type** value from **Slider** to **Rigid** using the dropdown. The parts will shake, showing that both are now connected. Click **OK** to finish the command:

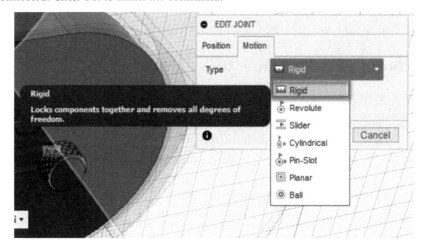

Figure 10.45 – Changing from Slider to Rigid joint

4. Click on **ASSEMBLE** and then click on **As-Built Joint**. The reason why we are doing an as-built joint is that holders 01 and 02 are in the locations that they should be in if they were assembled:

Figure 10.46 – Showing the location of As-Built Joint

5. Choose both parts and then choose the slot circle at the end:

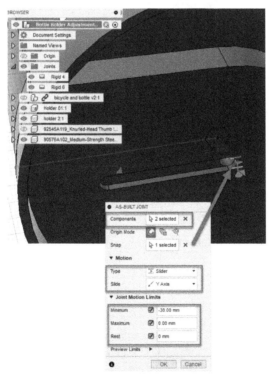

Figure 10.47 – As-Built Joint settings for holders 01 and 2

Now, when you move the Holder 2 body, the knob and hex nut will slide as they were meant to do:

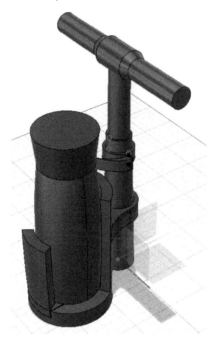

Figure 10.48 – Showing the completed bottle holder

Congratulations! You completed the bicycle holder. Remember—there is always room for improvement in your designs, and with the help of 3D printing, you can create a part quickly and see any errors, then go back into Fusion and readjust within just a few hours.

Now, we will place our parts on a drawing and add some dimensions so that we can share our design with other creators.

Creating a part drawing

Placing drawings on a sheet helps to show the size of a part and how multiple parts may join. To fit large-sized objects onto a small sheet, you set the part to a scale, which lets someone know that the part has been scaled down to fit onto a drawing sheet.

For this section, we will pull in the parts and place them on a few separate drawings. One drawing will show the entire assembly, the next will show Holder 01, and the third will be Holder 2. Normally, you would dimension every area that a fabricator would need to create this part, but since we will be 3D-printing it, we will place some basic overall size dimensions.

Let's get started:

1. Open your bottle design project, if you had previously closed it out, click on the **DESIGN** drop-down arrow, and go to **DRAWING | From Design**:

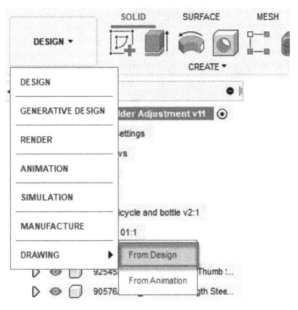

Figure 10.49 – Drawing workspace location

2. A **CREATE DRAWING** pop-up window will appear. Set the following selections:

    -   Under the **Reference** area, change the **Contents** dropdown's value to **FullAssembly**, which means it will grab everything even if the eyeball icon is turned off

    -   In the **Destination** area, set **Drawing** to **Create New** (the other option would be to add to an existing drawing if one was already made)

    -   The **Template** setting will be **From Scratch**, but if you have a company and want to use the company templates, you can use them here

    -   Set the **Standard** type to **ISO** since we created this part with metric units

    -   Set **Units** to **mm** but this can be changed to imperial units if needed

    -   Set **Sheet Size** to **A3 (420mm x 297mm)**, or for larger parts and assemblies, you can change this to whichever setting fits the part well

Once all of this is set, click on **OK**:

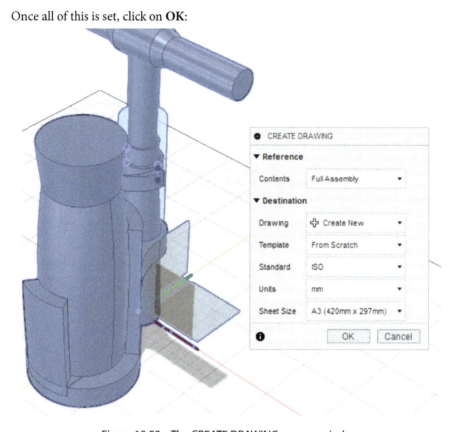

Figure 10.50 – The CREATE DRAWING pop-up window

A new tab will open in Fusion showing a drawing sheet, a popup, and the 3D model we created attached to your cursor. If you accidentally left-clicked and placed the part, no worries—we can always adjust or remove and place another part in. We won't go over everything, but the area to concentrate on first is the **DRAWING VIEW** area on the right side.

Left-click to place the **FRONT view** area of the part in the general location, as I have done in *Figure 10.51*:

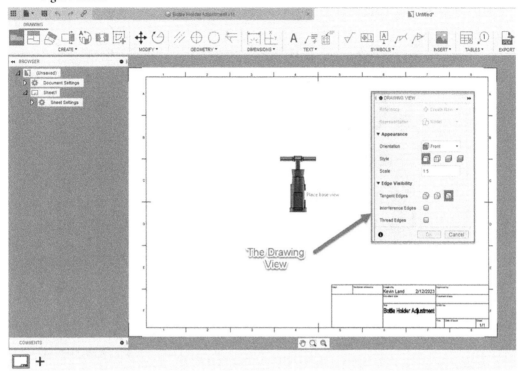

Figure 10.51 – The drawing workspace overall

3. Change the **Scale** value from 1:5 to 1:2, and notice that the size of the part grows. You set the scale of a part to fit the size of your paper. Typically, I leave the **Orientation** type as **Front** and create my other views from this base view. Click **OK** to accept the other defaults for now:

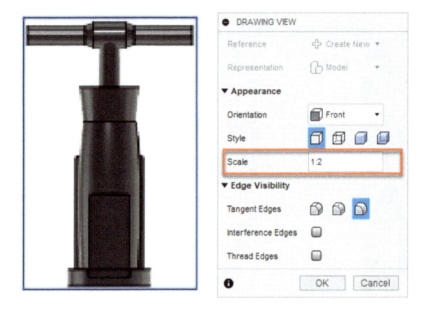

Figure 10.52 – The DRAWING VIEW flyout

Let's save our drawing before we get too far in.

4. Go to the top left and click on the **Save** icon.

Fusion will open a pop-up window that takes the original filename and adds the text **Drawing** at the end. It also places it in the current working project folder since this is a separate file; it is not contained within the current project. Click the **Save** button in the bottom right:

Figure 10.53 – Saving the drawing file

5. You can create more views by creating projected views from the first view that was placed on screen. Click on **Projected View** within the **CREATE** panel, then left-click on the **FRONT** view that you first placed and move your mouse to the right, then left-click, and hit *Enter*:

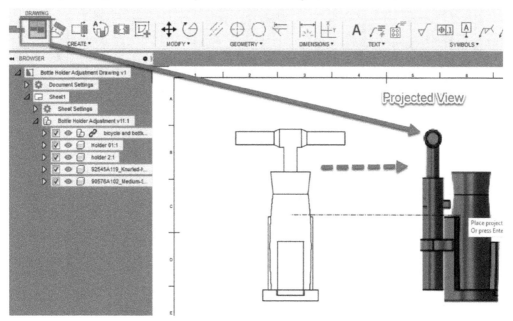

Figure 10.54 – Creating a right-projected view

We want to create more views, but we don't have any more room for our drawing. We can fix this by adjusting the paper size or just adjusting the view scales.

6. Double left-click on the **FRONT VIEW** that you first placed. The **DRAWING VIEW** window will appear again; click on the ellipses next to the **Scale** option, and you will see multiple scales appear. Change it to **1:4** and notice that both views change size. This is because the first view that is placed is the "master" view. If that "master" view changes, then all other attached views change as well:

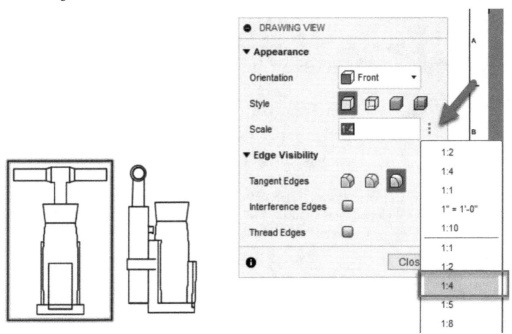

Figure 10.55 – Adjusting the scale

7. Left-click on **Projected View** again and create a **TOP** view. If you need to make space, you can left-click and drag on the **FRONT** view and move it down a bit on your drawing screen; this will move your right view as well:

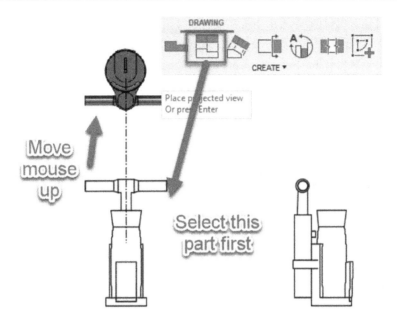

Figure 10.56 – Adding a top-projected view

Now, let's create a projected view by left-clicking on the **FRONT** view and then moving your mouse to the bottom left.

8. Left-click and hit *Enter* to place the view:

Figure 10.57 – Adding a projected view

You can change each individual view to different scales and styles.

9. Move the projected view by left-clicking and dragging and move it to the top right.

10. Double-click on the projected view, which will open the **DRAWING VIEW** window, and change the **SCALE** value from **1:4** (from parent) to **1:2** and change the **Style** type to **Shaded**. Select **CLOSE** to finish the command.

11. Most of the time, you keep the main views such as **TOP**, **SIDE**, and **FRONT** at the same scale and can set the projected view to a different scale since you normally don't place dimensions on it:

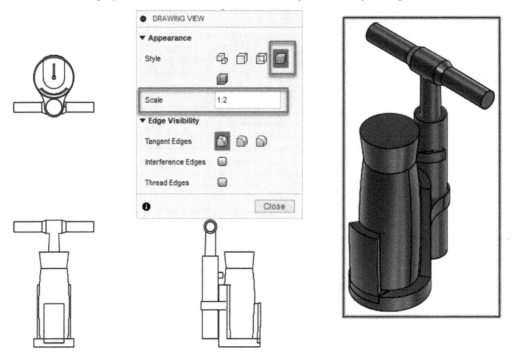

Figure 10.58 – Changing the projected view

12. To place text under each view, click on the **Text** button, then move your mouse under a view, and left-click to place a corner point. Then, move your mouse to the opposite corner and left-click once more to create a box. Type in some text that corresponds to the view, then left-click outside the box (green arrow) to complete the command:

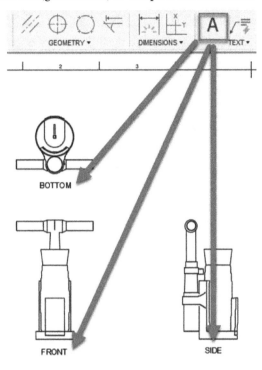

Figure 10.59 – Adding text

Next, let's create a **bill of materials** (**BOM**) that shows a person what each part is and where it is located.

Creating a BOM

Follow these steps to create a BOM:

1. A BOM will be automatically created and automatically bubble your items once you click on **TABLES | Table** and then left-click and place it into the bottom-right corner above the title block.

    Item bubbles will automatically be placed on the bottom view since this is where it sees the most items. Item number **5** is not shown as it is currently hidden under the water bottle. We will need to create another view to see it:

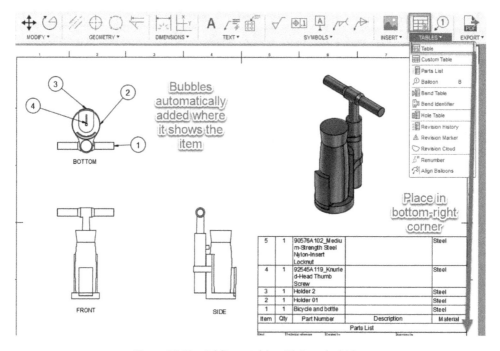

Figure 10.60 – Adding a table with item bubbles

2. In the bottom-left corner of your screen, right-click on the small drawing icon, then select **Duplicate Sheet**. This will open a copy of the previous drawing:

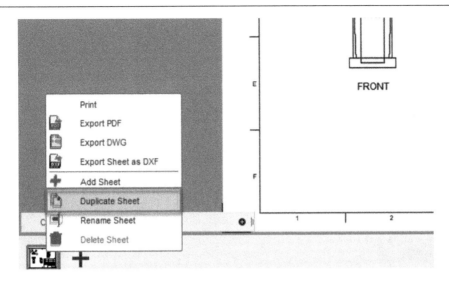

Figure 10.61 – Duplicating a sheet

3. Left-click on the check mark box for the bicycle and bottle within the **BROWSER** section.

This will turn off the 3D models for the handlebars and bottle, allowing us to see the hidden item number **5**. Unfortunately, we lost item number **1**, so now our numbers are **2** through **5**, which doesn't look good for a BOM:

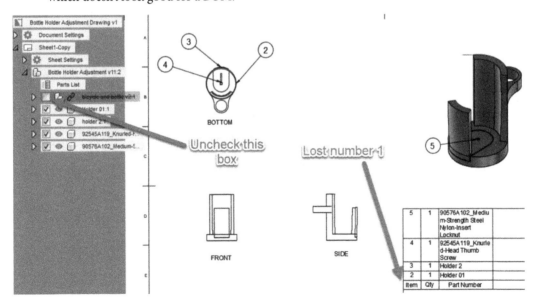

Figure 10.62 – Removing a 3D model

Typically, you only have a BOM on the first sheet and then add item numbers to the items only on that first sheet.

4. Let's remove the table on sheet **2** by left-clicking on it and hitting *Delete*, which will remove the bubbles and table. Now, turn off Holder **2**, plus the bolt and nut. Double-click on the **FRONT** view and change the **Scale** value to **1:2** since we have more sheet space than before:

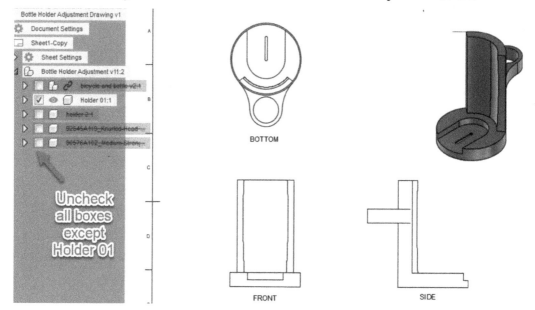

Figure 10.63 – Removing the table and bubbles and extra items

Let's create another sheet for Holder 2.

Creating a sheet for Holder 2

Follow these steps:

1. In the bottom-left corner of your screen, right-click on the sheet we are currently on and duplicate it.

2. In the **BROWSER** section, remove the check mark for Holder 01 and add in a check for Holder 2.

3. Double-click on the **FRONT** view and set it to a **1:1** scale since this is a smaller piece:

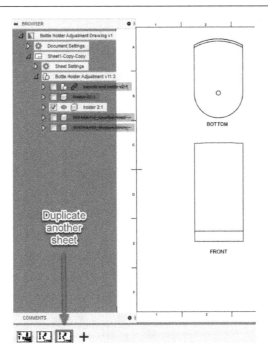

Figure 10.64 – Duplicating another sheet

Let's remove the **BOTTOM** view and create a sectional view through the side view looking down.

4. Left-click on the **BOTTOM** view and hit the *Delete* key on your keyboard. Remove the bottom text as well by left-clicking and deleting. Click on the **FRONT** view and move it up on the drawing along with the text:

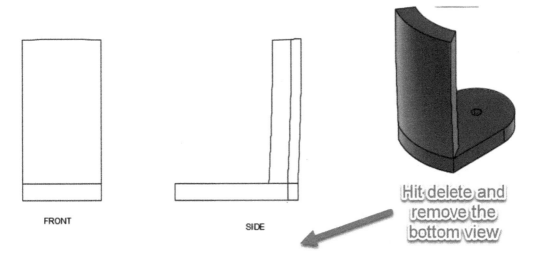

Figure 10.65 – Removing a view

Now, let's create a section view looking down at the top of the side view. A section view creates a view looking from wherever the section line is placed. This helps to locate parts that may be hidden or view depths of holes that may not go all the way through a part.

5. Click on the **Section View** button in the **CREATE** panel and then left-click on the **SIDE** view. A letter **A** will appear where your cursor is.

6. Left-click once to place an arrow on one side (step **1** in *Figure 10.66*) and then move your mouse to the right (step **2** in *Figure 10.66*) and click to place the other arrow.

7. Hit *Enter* to complete placing the arrows. Now, click **OK** in the **DRAWING VIEW** popup:

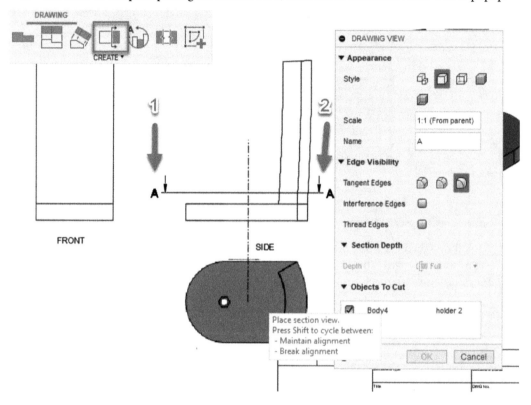

Figure 10.66 – Adding a section view

Now that we were able to create our views, let's add some dimensions so that we know what size of the part we will be using.

Adding dimensions

Let's add some dimensions to give the general size of these parts. Typically, these would be for a machinist to create this part, but since we will be 3D-printing, we won't worry too much about adding all dimensions. Let's get started:

1. Click on the **Dimension** icon, which looks like a little sun.

    This sun icon means that this is a smart dimension, and when it is activated, if you mouse over lines, circles, or arcs, it will automatically guess which kind of dimension you want to place, such as linear, diameter, aligned, and so on. With the "sun" dimension on, mouse over the bottom line of the front view.

2. Drag your mouse down a little bit and left-click to place the dimension, then left-click again to place its location:

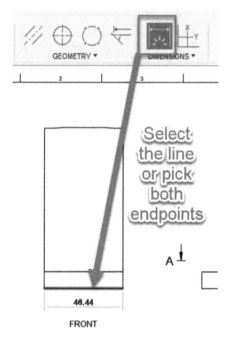

Figure 10.67 – The Dimension tool

3. Click on the **DIMENSIONS** drop-down arrow and select **Linear Dimension**.

4. Left-click on the top center to see the **Quadrant snap** icon and left-click, then click on the bottom-left corner to select that endpoint.

5. Now, move your mouse to the left, and click to place that dimension:

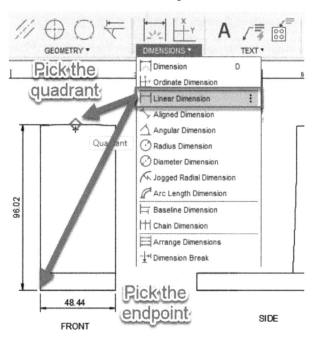

Figure 10.68 – Adding a linear dimension

6. Click on either the "sun" dimension or the **Radius Dimension** setting and select the circle. Left-click to place the dimension:

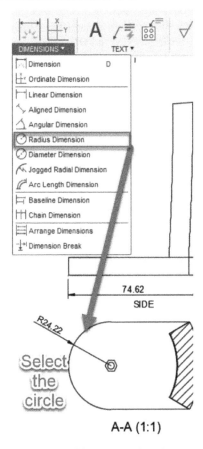

Figure 10.69 – Adding in a Radius dimension

Continue adding in more dimensions for this sheet and the other three sheets until you feel that the overall sizes have been added. If we had used the **Hole** command to create the thumbscrew hole, we could have used the **Hole and Thread Note** tool to show the size of the hole with proper noting.

Now that we placed our dimensions, what happens when we need to make a change to the model? That is no problem for Fusion 360—we can make a minor change, and the model will adjust and keep the dimensions. Be aware, though, that making major changes may result in dimensions becoming disconnected.

Making an adjustment to the 3D model

If you make any adjustments to the original 3D model, your drawing will need to be refreshed to see those updates. Let's add in a fillet to Holder 01 and then update the drawing, as follows:

1. Open the **Bottle Holder Adjustment** model.

2. Turn off all models except for Holder 01 (green arrow). Then, select the **Fillet** tool and left-click on the top and bottom corner curves (red arrows), and set the distance to **10** mm. Click **OK** to finish the command:

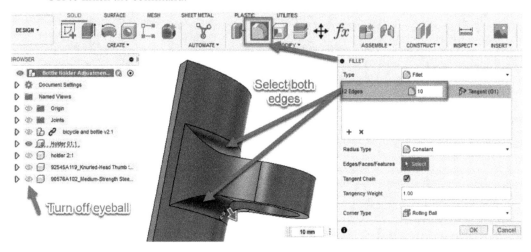

Figure 10.70 – Adding in fillets

3. Save your changes and add a note of the changes that were added:

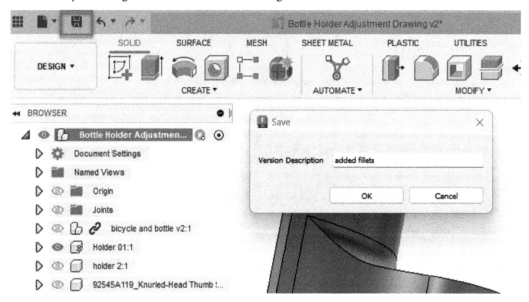

Figure 10.71 – Saving the changes

4. Go back into your drawing file and note that a small warning icon appears in the bottom-right corner. Go to the top-left corner and click on the warning icon:

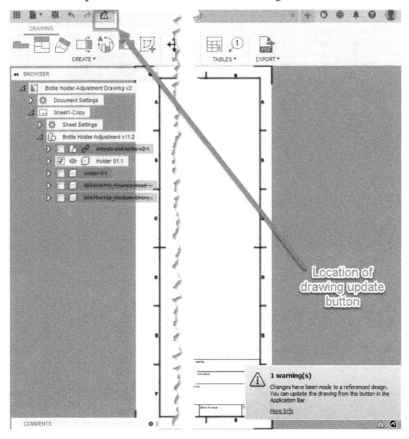

Figure 10.72 – Warning icon and message

The drawing updates with the newly added fillets. Be aware that some dimensions may update if a size changes, but some may lose connection, which means you may have to redo the dimensions for your drawing file:

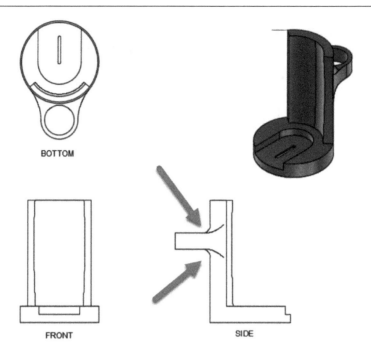

Figure 10.73 – Updated drawing file with new fillets

There is a lot more to cover in the sheet workspace but I will leave that up to you to explore. Create, make mistakes, and try again! That is the best way to learn.

Let's continue with more practice models!

Summary

In this chapter, we were able to finish off the bottle holder for our bicycle! At the beginning of this chapter, we adjusted the model with a few changes using the **Press Pull** tool and removed material using the **Shell** tool. We then added parts from the McMaster-Carr online part library and were able to set them with joints. We then finished off the chapter by creating a part drawing with dimensions for our model and then adjusted the model one more time to see the drawing views change to fit the adjustments.

In the next chapter, we will start a completely new project and create a model using the **Form** environment.

Part 3: FORM Modeling Techniques

In this part, we are introduced to the form environment, which is non-parametric, meaning that we are free to create without restrictions. We will first be introduced to how to the tools work and how to shape simple base objects into workable parts. We will then create a simple tealight ghost and learn how to use the form environment to create a gaming chair.

This part has the following chapters:

- *Chapter 11, The FORM Environment*
- *Chapter 12, Modeling a Scary Tealight Ghost*
- *Chapter 13, Using Form and Solid Modeling to Create a Cushioned Chair*

11
The FORM Environment

In this chapter, we will learn how to create mesh objects within the **FORM** environment. The **FORM** environment doesn't use any parametric tools or constraints, which means you are able to freely adjust your model as you would if you were modeling with a block of clay.

We will first learn how some of the tools work by experimenting on a box model and, once we are familiar with the typical tools used within the **FORM** environment, we will create an organic computer mouse. By the end of this chapter, you will know how to use the **Edit Form** tool, create form models, and edit these types of models.

In this chapter, we're going to cover the following main topics:

- Overview of the form environment
- The **CREATE** and **MODIFY** tools
- The **Edit Form** tool

Technical requirements

You can practice with the files provided or feel free to create your own for a more custom experience. The sample design for this chapter can be found at `https://github.com/PacktPublishing/Improving-CAD-Designs-with-Autodesk-Fusion-360/tree/main/Ch11`.

Overview of the FORM environment

We are going to switch gears from the parametric design space now and start working within the **FORM** environment. The **FORM** environment is another modeling space that does not use any parametric tools to create models. There is no history bar at the bottom of your screen, meaning that all the mistakes or changes that you make will have to be manually changed or fixed with the **Undo** command.

We will go over a brief overview of the environment, where the most common tools are located, and how to use them in this section.

To get to the **FORM** environment, you must go to the **CREATE** panel and click on the **Create Form** tool.

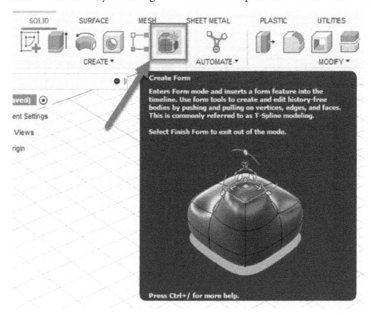

Figure 11.1 – The location of the Create Form tool

Your workspace will change and show new tools at the top of your screen. The **CREATE** and **MODIFY** tools have changed and there are now tools such as **SYMMETRY** and **UTILITIES**.

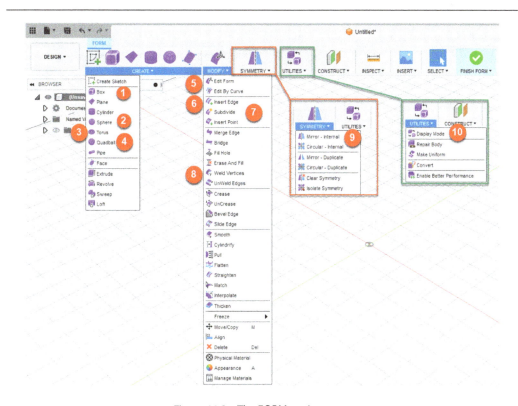

Figure 11.2 – The FORM environment

The most typically used tools are listed in the following table.

Bubble #	Icon	Name	Description
1		**Box**	Creates a Form Box using T-Splines.
2		**Sphere**	Creates a T-Spline Sphere.
3		**Torus**	Creates a donut-shaped T-Spline.
4		**Quadball**	Similar to a Sphere, it creates a rounded cube shape using T-Splines.
5		**Edit Form**	One of the most important and most used FORM tools. It can move, rotate, and scale points, edges, and faces.

Bubble #	Icon	Name	Description
6		**Insert Edge**	Adds another edge to an existing T-Spline body.
7		**Subdivide**	Splits a single T-Spline into multiple faces.
8		**Weld Vertices**	Connects two endpoints together.
9		**Mirror Internal**	Creates a symmetrical edge line, which results in only having to work on one side of a model.
10		**Display Mode**	Changes from box mode to smooth mode, which helps to see whether edges or points are interfering with other locations.

Table 11.1 – Most used FORM environment tools

Most of the time, when working with organic shapes, you try to start with a shape that looks like what you are trying to model. If you've worked with Blender, you may know of a well-known YouTube tutorial on "how to create a donut," which starts with a torus shape, a geometric shape that has been in many 3D modeling tools for many years.

You may have noticed that the word **T-Spline** appears many times in the preceding table. So, what are **T-Spline**s, you may ask? A **T-Spline** is a type of modeling where a point can be removed without destroying the model shape. For instance, in *Figure 11.4*, I have a **Box T-Spline** model with a highlighted point on the left side. If I hit *delete* and remove the dot, the geometry does not create a hole but creates a T-shape, allowing the model to not break, as shown on the right side.

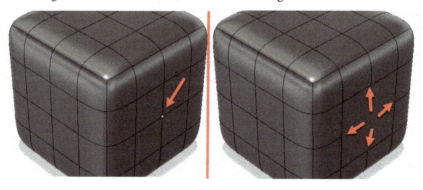

Figure 11.3 – Example of a T-Spline

Now that we have a better understanding of some of the tools and where they are located, let's start working with form models.

Working with form models

One of the main things to know about working in the **FORM** environment and with form models is that you can create sketches the same as in the **Design** environment, but sketches do not work the same. The sketches in the Form environment are not parametrically connected to the solid model but can be used to create a general size of a model. The following is an example of this, so please try this out for yourself:

1. Set **Units** to **Millimeters** by going to **BROWSER** on the left, then **Document Settings**, and then **Units**.

2. Create a sketch on the bottom plane of a rectangle with the dimensions 100 mm x 50 mm.

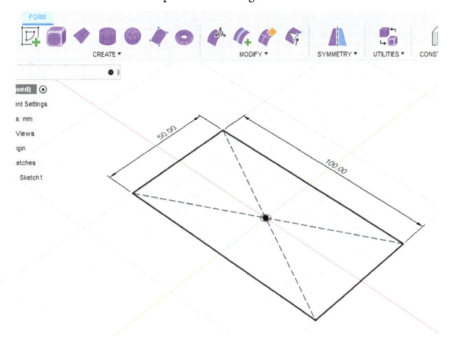

Figure 11.4 – Showing a sketch within the FORM environment

3. Click on the **CREATE** panel dropdown and then click on **Extrude**.

4.   Set **Distance** to 50 mm and click **OK**.

Notice that the extrusion preview has no top or bottom. It also has very thin walls for the body. This is a **T-Spline** body with quad faces.

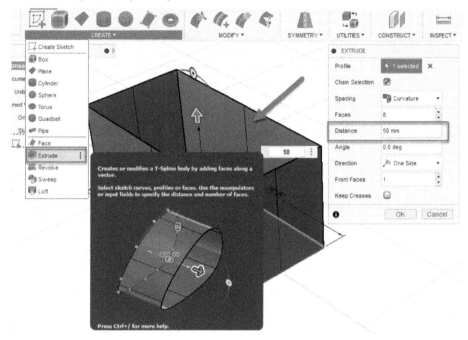

Figure 11.5 – The Extrude FORM tool

5.   Turn on the **Eyeball** icon for the sketch that you just created and double-click on the **Sketch** icon to edit the sketch.

6. Change the dimensions from 100 mm x 50 mm to 200 mm x 200 mm.

    Notice that nothing changed for the extruded model, but the sketch did change size. The sketches in this environment can help you create a specific-sized object to start with, but they will not adjust since this environment is used for freeform modeling.

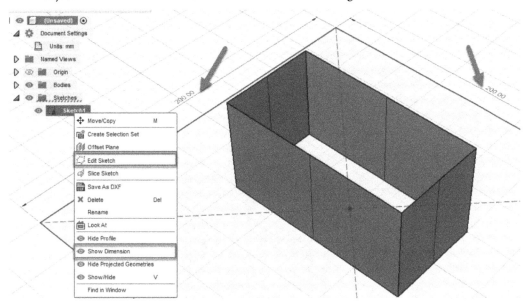

Figure 11.6 – The Sketch tool not changing size

Important note

To show dimensions when you are outside of a sketch, right-click on the sketch name, and then select **Show Dimension** as noted in *Figure 11.6*.

Next, let's fill in the hole for the top and bottom of the rectangle that we just created.

Filling in a hole in the FORM environment

There are two possible ways to fill in any gaps or holes in the **FORM** environment. One way is by using the surface tool **Fill**, which we will discuss in *Chapter 12*, and the other is by using the **Fill Hole** tool, which we will go over now:

1. Go to the **MODIFY** drop-down arrow and select the **Fill Hole** tool.

Figure 11.7 – The Fill Hole tool's location

2. Click on the top edge of the box and notice that the hole collapses down since the first **Fill Hole Mode** option is **Reduced Star**. If you cycle through each drop-down option, you'll notice that each one has a different fill.

> **Important note**
>
> We will not go into the weeds about how each fill mode creates its fill but just know that it is dependent on how the hole was created, meaning you always want to try to reduce the number of faces and points as much as possible and not make the geometry too complicated.

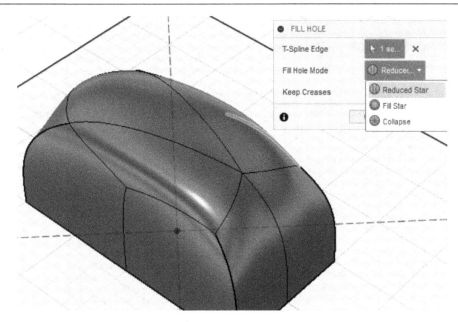

Figure 11.8 – The FILL HOLE tool options panel

3. Select the **Collapse** option, then click the checkbox next to **Keep Creases**. This will keep the edges and fill the hole without smoothing out the edges.

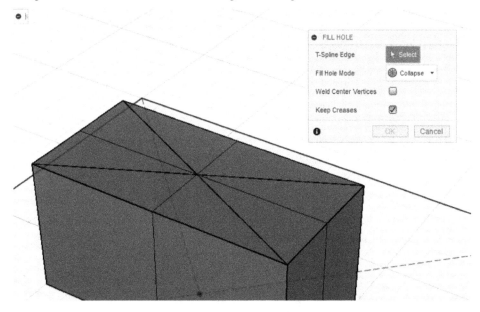

Figure 11.9 – The Fill Hole tool with Keep Creases checked

4. Before repeating the process for the bottom hole, click on **FINISH FORM**. This brings you back into the **DESIGN** environment.

5. Turn off the sketch by clicking on the eyeball and open the `Bodies` folder.

    Notice that the icon is not the typical body icon that we have been used to. This new body is called a **Surface**. Surface bodies are thin models without any mass. Think of them as tin foil bodies that you can mold and shape, but they have no thickness.

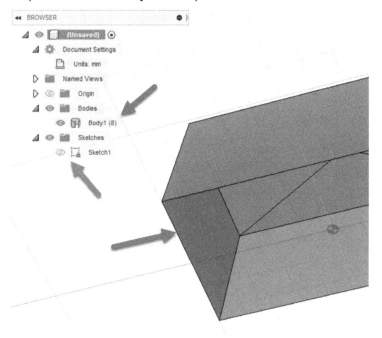

Figure 11.10 – The Surface body

6. Go to the timeline at the bottom of the screen and double-click or right-click on the **Form** icon and then select **Edit** to go back into the **FORM** environment.

7. Now fill in the second hole with the same options as before and then click on **FINISH FORM**.

8. When you return to the **DESIGN** environment this time, the `Bodies` folder will now show the typical solid body. The reason for this is that to create a solid body from the **FORM** environment, it will need to be fully enclosed with no holes; otherwise, you will be creating a surface.

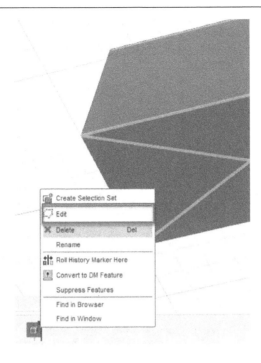

Figure 11.11 – Edit the form to go back into the Form environment

Now that you have some basic knowledge of the **FORM** environment., we can get a bit more creative and dive deeper into it.

The CREATE and MODIFY tools

It's time to flex your artistic muscle now. In this section, we will explore the tools located within the **CREATE** panel and the **MODIFY** panel. These are the main areas that you will spend most of your time creating in within the **FORM** environment.

Let's start by creating some artistic forms by using the **CREATE** and **MODIFY** tools together:

1. Start a new drawing by clicking on **File | New Design** or by clicking on the plus (+) icon in the top-right corner.

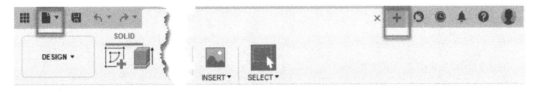

Figure 11.12 – New Design location

2. Click on the **FORM** tool within the **CREATE** panel to be brought into the **FORM** environment and then click on the **Box** tool within the **CREATE** panel.

Figure 11.13 – The location of the Box tool

3. Select the bottom plane, then choose the center origin dot, and in the **BOX** options on the right side of the screen, choose **Center** for the **Rectangle** option.

> **Important note**
>
> Be aware that once you create your rectangle, do not hit *Enter* on your keyboard or you will place a box without any further customization options.

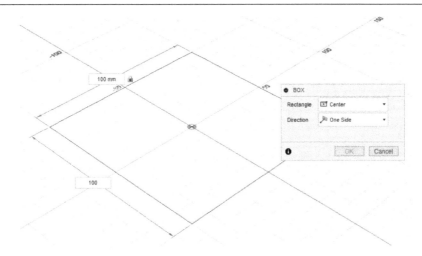

Figure 11.14 – The BOX tool placement options

4. Set **Length**, **Width**, and **Height** to 100. Notice that the cube has, by default, smoothed surfaces on all edges.

Important note

Do not close out of the box window or click **OK**. If you do, please create another Box, as you cannot get these options back unless you create another box.

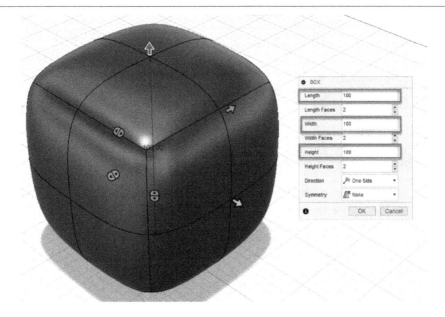

Figure 11.15 – The FORM tool BOX settings

5.   Increase **Length Faces**, **Width Faces**, and **Height Faces** from **2** to **4**. This will start to square off the cube. We could keep increasing the edges, but this would make the model almost unworkable due to all the faces. It's best to keep the face count low since we can always add more if needed as we work.

6.   Click **OK** to complete the command. I usually create enough to get the rough shape and have a centerline on each face, in case I would like to mirror a side for symmetry.

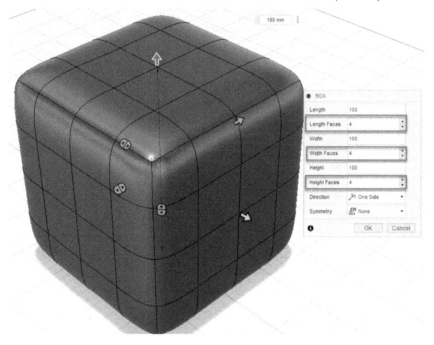

Figure 11.16 – Increasing the face counts

Now that the base feature has been created, let's learn how to select faces, edges, and points and manipulate them to change the geometry.

Selecting features on an object

Learning how to move and manipulate features is the key to creating a decent model in the **FORM** environment. Let's start by selecting some faces, edges, and points:

1.   Select a face somewhere around the middle of the cube by clicking on a square. If you are unable to select a face, edge, or point, you may want to check your selection filters and make sure that they are on (see *Figure 11.17*).

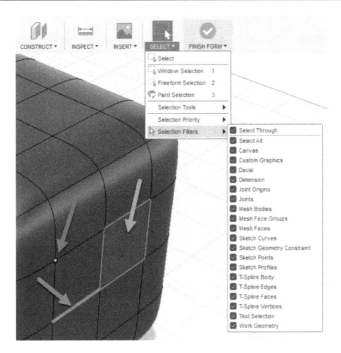

Figure 11.17 – Selection filters all on

2. If you left-click and drag, you can select multiple objects, but it may select faces that you may not want to select on the opposite side. To only select faces on the side that you want, you can either hold down the *Shift* key and select each object or you can uncheck **Select Through** within the **Selection Filters** dropdown (see *Figure 11.18*) to only select faces on the visible side.

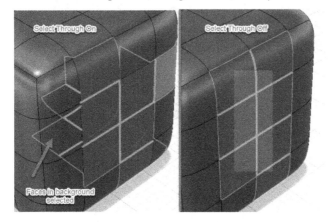

Figure 11.18 – Select Through option

Now that we know how to create and adjust a box, the next section will lead us through how to edit the shape.

Filling holes

Filling holes in poorly made geometry is something that all 3D modelers will have to do. Sometimes, it can be easy and sometimes it can be a pain. Follow these steps to learn how to fill a hole in a 3D model:

1. Select the four inner faces and hit the *delete* key.

    Notice that it creates a hole since the faces have now been removed while, earlier in the chapter, we removed an edge that did not result in a hole but a **T-Spline**. Also notice that the box is hollow, just like the sketch we created earlier.

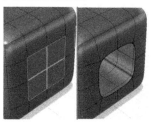

Figure 11.19 – Hitting delete with faces selected

To fill the hole, we could use the **Fill Hole** tool as before, but the options within the tool, as you may have noticed, do not result in the same simple square geometry that was there before. So instead, we can use the **Bridge** tool.

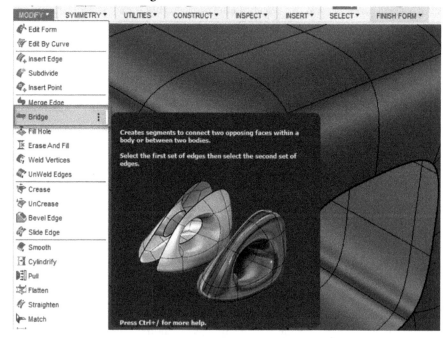

Figure 11.20 – Bridge tool location

2. Go to the **MODIFY** panel dropdown and select the **Bridge** tool.

3. Select **Side 1** by holding down the *Shift* key and left-clicking on the top two edges. Similarly, choose **Side 2** by selecting the bottom two edges.

4. Select the **Preview** checkbox to see the selection and set **Faces** to **2** since we want to add two faces to fill in the hole. We could add more faces if we wanted, but we always want to try to have clean geometry when working with **FORM** models. Click **OK** to complete the command.

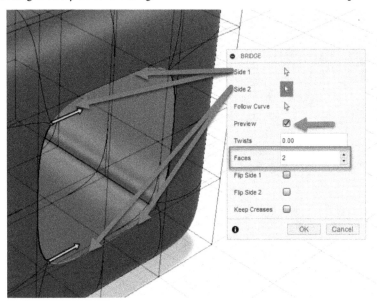

Figure 11.21 – The BRIDGE tool options

The hole has been filled but something looks wrong with the lines. There are some odd waves within the geometry. To fix this, we need to use the **Weld Vertices** tool.

Figure 11.22 – The hole is filled but there are points not welded

The points along the side were not connected like the points at the top and bottom. We need to manually fix this by using the **Weld Vertices** tool located in the **MODIFY** panel dropdown.

5.   Activate the tool and you can pick points (use a window selection when the points overlap, or you can pick each individually) to join them back together (shown in red circles). Notice that the dark black lines (orange arrows) are a visual cue that there is something wrong with the geometry.

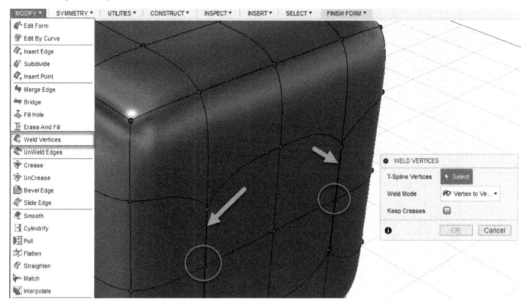

Figure 11.23 – The Weld Vertices tool location

Now that you know how to fix holes and weld them back together, let's create a simple shape using the **Edit Form** tool.

The Edit Form tool

The **Edit Form** tool is one of the main tools that you will be using within the **FORM** environment since it can do a variety of things, such as scale, move, and rotate, all within the same tool. Here is a list that displays the different tools we will use:

a.   Edit Form tool location

b.   Edit Form tool window

c.   Set Pivot

d.   Scale in all directions

e.   Scale along axis

 f. Move along axis

 g. Move along plane

 h. Rotate handles

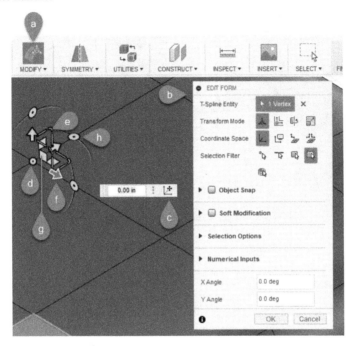

Figure 11.24 – The Edit Form tool information

One of the most important things when using the **Edit Form** tool is to be aware that you should not create self-intersecting geometry. This means do not overlap faces, points, or edges. If this happens, then Fusion 360 will display a warning that you need to find and fix these errors or it will not create a solid 3D model. We will go over this next.

Fixing a broken mesh body

There are times when the mesh body may look correct in the **Create Form** environment but when you click the **FINISH FORM** button, Fusion 360 will display an error stating that there are "self-intersects." Luckily, Fusion 360 will show you a red line where the issues are happening so that you can adjust the model and fix it.

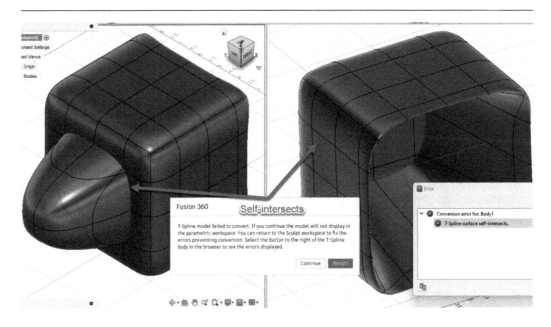

Figure 11.25 – Self-intersect warning

Follow these steps to cause and fix a self-intersect issue:

1. Select the **Multiple Views** tool at the bottom of the screen to change your workspace into four separate viewports. This will help us view all four sides of an object without having to pan and orbit around the model multiple times. You still have the ability to manipulate each viewport separately though.

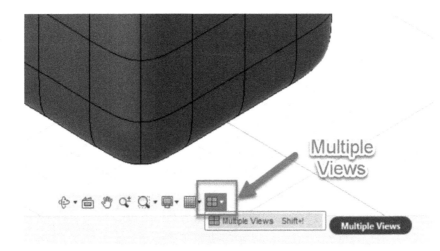

Figure 11.26 – Multiple Views icon location

2. Select four faces on the cube in the **Perspective** viewport (it's the angled view at the top right) and select the **Edit Form** tool.

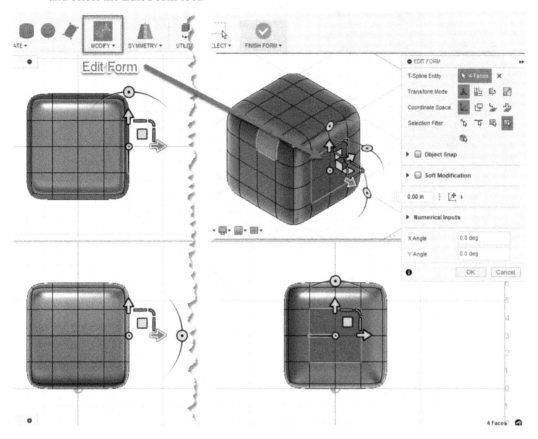

Figure 11.27 – Selecting four faces and using Edit Form

3. Drag the X Distance arrow about 8.5 units, to the opposite side of the cube. Click **OK** to finish with the **Edit Form** tool.

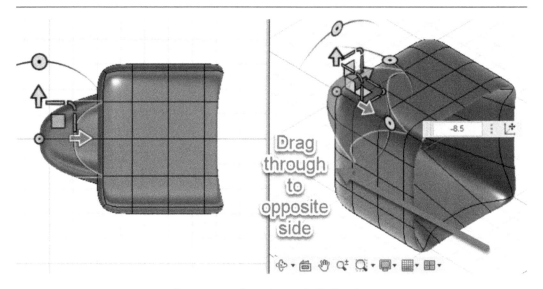

Figure 11.28 – Dragging in the X direction

4. Click on the **FINISH FORM** button and notice that Fusion 360 pops up a warning about self-intersects and that it failed to convert the mesh into a solid body (see *Figure 11.25*). If you click on the **Continue** button, you will be brought back into the Create environment, but your model will disappear due to Fusion 360 not knowing how to convert the mesh into a solid body. Click on the **Return** button to be brought back to continue working in the mesh environment.

5. There are two ways to fix this issue, but both options are that you need to separate the overlapping bodies so that Fusion 360 can run the conversion. You can do this by either trimming/deleting the extra material away or using the **Edit Form** tool to drag faces backward, away from the intersection:

    a. Select the same four faces as before by orbiting a viewport to see the front side of the box and clicking on each face.

    b. Drag in the X direction about 3 units back.

    c. Click **OK** to finish with the **Edit Form** tool.

    d. Click on **FINISH FORM** to create a solid body.

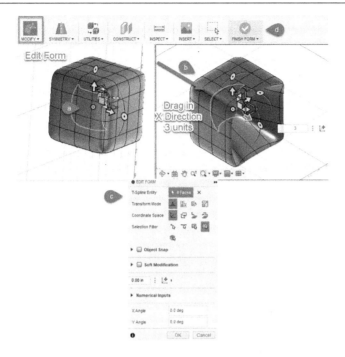

Figure 11.29 – Dragging away to fix a self-intersect

6. Click on the **Viewports** tool and select **Single View** to bring your multiple Fusion viewports back to a single view.

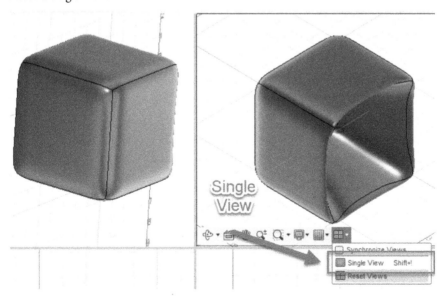

Figure 11.30 – Selecting the Single View option

This is an issue that many organic modelers (people who work with organic shapes such as humans, animals, etc.) possibly run into while working. Now that you know about this issue, you can avoid this mistake, and if it happens by accident, you know how to fix it. Let's continue with another exercise on creating an ergonomic mouse.

To get a good idea of how to use this tool, we will create an ergonomic mouse, as pictured in the following screenshot.

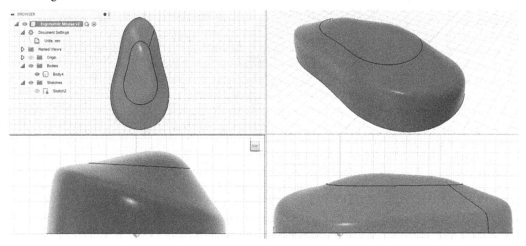

Figure 11.31 – The final FORM tool mouse result

Let's get started with the steps:

1. Start a new Design file and set **Units** to **Millimeters**.

2. Create a sketch on the bottom plane and create two circles with lines that are tangent on either side and a center construction line down the middle connecting the two center points.

3. Go to **MODIFY** and use the **Trim** tool to remove the two inner circle edges so that it matches *Figure 11.32*.

4. Set the dimensions as shown in *Figure 11.32*. These dimensions are mainly to get our model to around the size of the real-world ergonomic mouse.

5. Click on the **FINISH SKETCH** button.

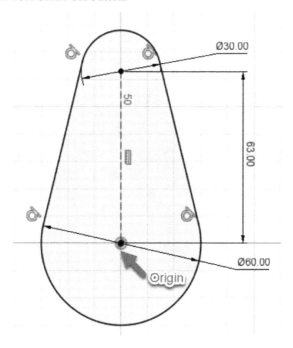

Figure 11.32 – Sketch of ergonomic mouse bottom with dimensions

6. Click on the **Form** tool and then click on the **EXTRUDE** tool within the **CREATE** panel and set **Distance** to 6 mm.

7. Set **Spacing** to **Uniform** since we want a nice centerline to be at the top of the circle and at the center of the bottom circle as well.

8. Set **Faces** to **12** and notice that as we increase the number from **8**, it starts to smooth out the circle a bit more.

Please experiment with lower numbers to see what happens to the shape, and with higher numbers as well. Your computer may slow down with higher numbers, which is the reason why we use lower numbers to start. Set the Distance to 6 and click **OK**.

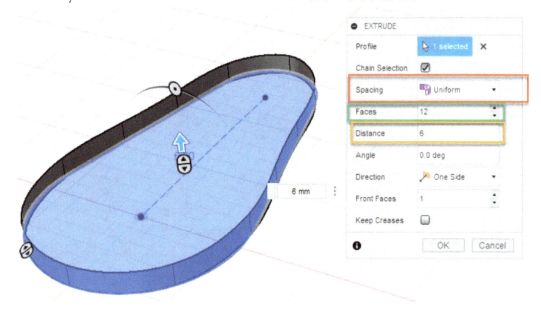

Figure 11.33 – The FORM EXTRUDE tool options

Important note

Notice that you can start the **Sketch** tool within the **Design** environment or the **FORM** environment since the sketch can live in both places.

---

9. Click on the **Edit Form** tool within the **MODIFY** panel and select a point along the top edge. Notice that some arrows appear as well as some lines, circles, and a few squares:

- The arrows are for moving
- The lines are for scaling along an axis
- The center dot is for uniform scaling
- The circles are for rotating
- The squares are for moving along an axis such as XY, YZ, or XZ

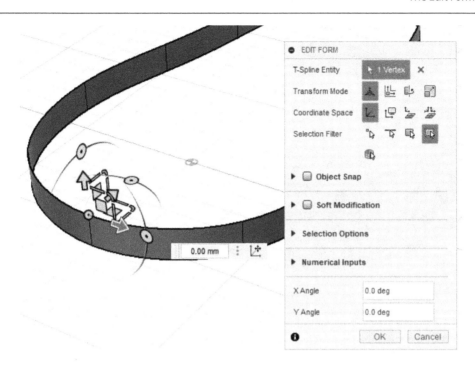

Figure 11.34 – The EDIT FORM tool

10. Select the arrow that is pointing in the Z direction and left-click and drag it up about 16 mm. Notice that the point that was selected moves along a straight path in the Z direction and the edges around it also move, maintaining a smooth curve.

11. Left-click to place the point. Notice that when you left-click to place the location of the point, you are still within the **EDIT FORM** tool. This is because the **EDIT FORM** tool allows you to modify so much that it would slow you down if you had to click **OK** every time you made an adjustment.

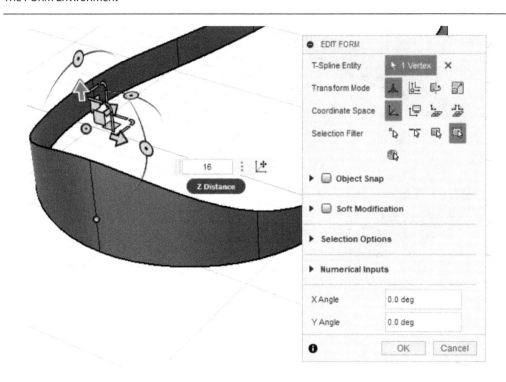

Figure 11.35 – Moving along the Z axis

12. Left-click on the same dot once again and notice that if you click on the same Z-direction arrow, you start at 0 mm instead of 16 mm.

13. Open the **Numerical Inputs** flyout arrow within the **EDIT FORM** tool flyout on the right and notice that all the options are set to **0** except for all the scale options, which are at **1**.

   The 16 mm movement does not show up anywhere in this box. The reason is that after a movement is complete, it treats lines, faces, and points as set at a new location and does not remember the last movement.

14. Set **Z Distance** to -16 and the point will move back down. If you want to set specific distances, you must remember them.

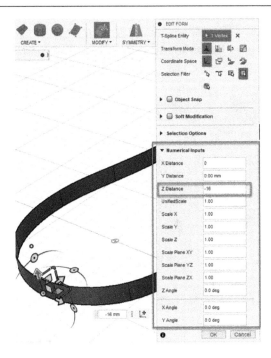

Figure 11.36 – The Numerical Inputs section

15. Double-click the top edge (the red arrow in *Figure 11.37*) and notice that the entire top ring is selected.

    This is a quick way of selecting multiple edges at once. We will move the pivot point of the **EDIT FORM** tool by setting a new location for it.

16. Click on the **Set Pivot** icon located next to the key-in distance floating window (green arrow).

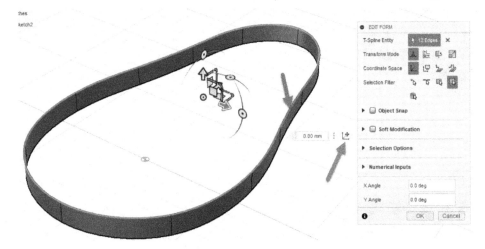

Figure 11.37 – Selecting an entire ring with a double left-click

17. Now click on a new point location (see the red arrow in *Figure 11.38*) and be sure to click on the green check mark (green arrow) or the new origin will not be set.

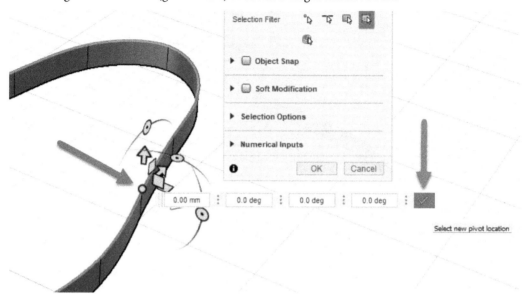

Figure 11.38 – Setting a new pivot location with Set Pivot

18. Click on the **Y Angle** rotation circle on the right to rotate the entire ring edge by 15 degrees. Left-click outside to place the rotation.

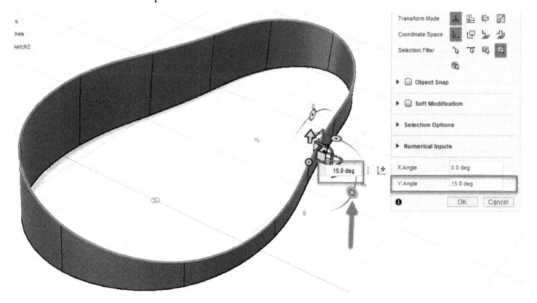

Figure 11.39 – Rotate the edge by 15 degrees

We will add some more faces since we will need more material to work with.

19. Double-click on the top edge again to select all the edges.

20. Now hold down the *Alt* keyboard key and click on the Z arrow and drag in the Z direction 2 mm. Left-click to place this new edge. You can add more edges and faces by holding down *Alt* and dragging in any direction.

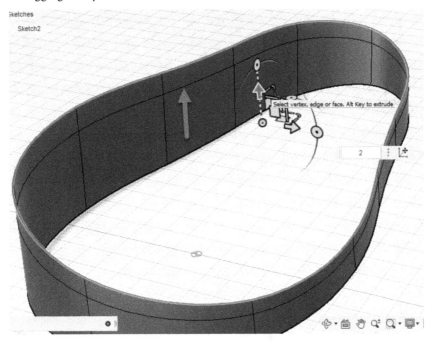

Figure 11.40 – Extrude faces using Alt and drag

We will scale the new edge inward to gain a curve at that location.

21. Double-click the new edge and left-click on the universal scale dot. Drag inward about **.9** units.

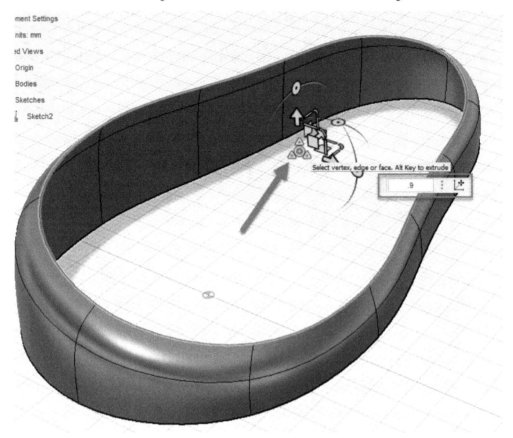

Figure 11.41 – Universal scale tool location

Let's now add another edge while scaling inward.

22. Once again, double-click the edge, hold down *Alt*, and select the **Universal Scale** tool and drag inward about **.7** units.

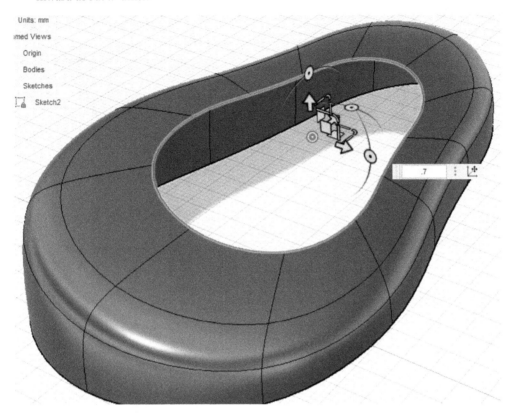

Figure 11.42 – Universal scale with Alt and drag

It's best to see your model from multiple views to make sure that you're not pulling points to another location.

23. Click on the **Multiple Views** button at the bottom of the screen to see a **Top, Front, Side, and Perspective** view.

24. Now double-click on the top edge and move the edge up about 7 mm and then rotate it -12 deg to level it off.

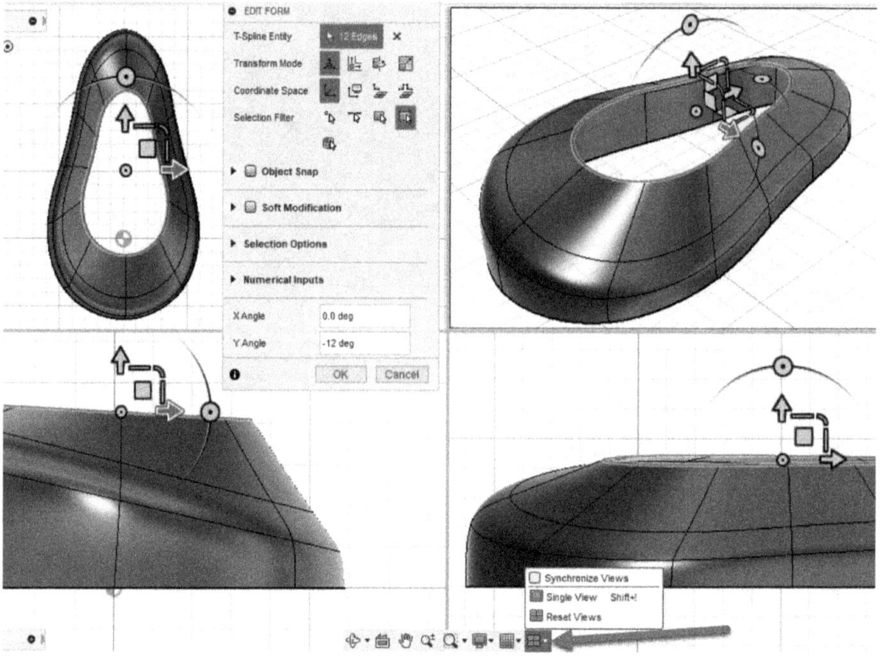

Figure 11.43 – The Multiple Views tool location

To close the top and bottom gaps, we will use the **Surface Environment** tool instead of the **Fill Hole** tool since the geometry created using that tool will be much cleaner.

25. Click **OK** to close out of the **EDIT FORM** tool and click the **FINISH FORM** button.

Now that we have been able to get the mouse body put together, we will use a surface fill tool to finish it off. You will learn while using Fusion 360 that, sometimes, a tool may not work the way that you want it to work, and you will need to lean on other knowledge of other tools that may help do the job.

Surface tools

The surface tools are much like the form tools in that they work only with thin, non-solid geometry. There are many tools to use in this tab, but we will only use the **PATCH** tool for now and we will explore more in *Chapter 12*:

1. Notice that we have a surface body since we did not close out the holes at the top and bottom. Click the **Single View** icon at the bottom of the screen to return to a single view from the multiple views.

2. Click on the **SURFACE** tab at the top of the screen.

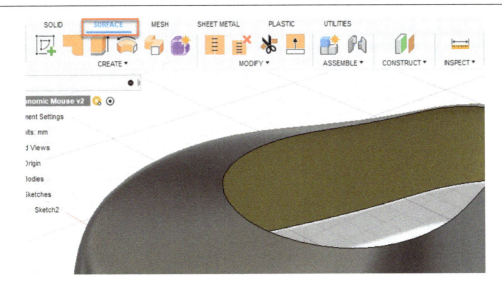

Figure 11.44 – The SURFACE tab location

3. Click on the **CREATE** drop-down arrow and click on the **Patch** tool.

Figure 11.45 – The Patch tool location

4. Orbit to view the bottom of the mouse and click on the edge. A preview of the flattened bottom will appear. Click **OK** to close the command.

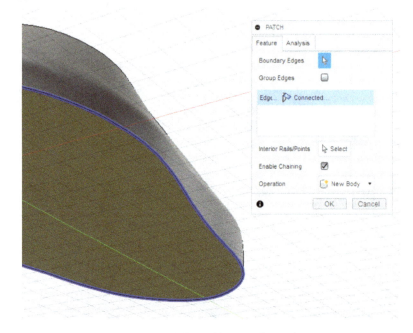

Figure 11.46 – PATCH tool bottom face selection

5. Click on the **PATCH** tool once again and click on the top-hole edge. Notice that it creates a flat patch like the bottom hole, but we would like it to be curved.

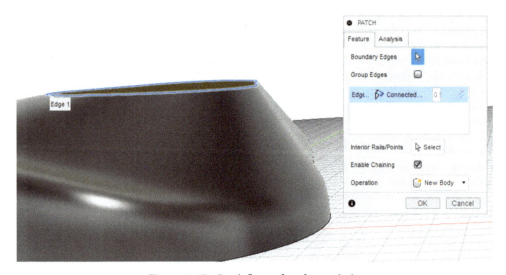

Figure 11.47 – Patch flat surface for top hole

6. Choose the **Tangent (G1)** curvature and notice that Fusion 360 maintains tangency to produce a curved top. Click **OK** to complete the command.

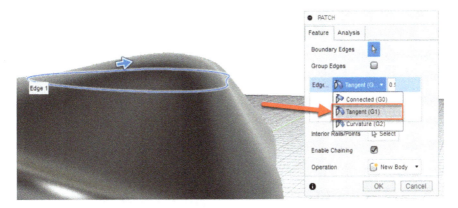

Figure 11.48 – PATCH tool tangent edge selection

Orbit the body to view the bottom once again and notice that the bottom patch that was created is a different color. We need to "reverse the normal direction" of the bottom patch so that if a material is applied, it will face in the right direction.

7. Go to the **MODIFY** dropdown and click on the **Reverse Normal** tool. Left-click on the bottom face and click **OK** to close the command.

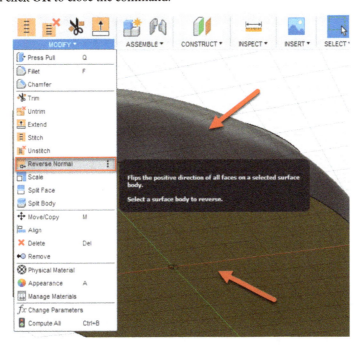

Figure 11.49 – The Reverse Normal tool location

The final step is to combine all the surface bodies together.

8.   Open the `Bodies` folder located within **BROWSER** on the left side of the screen and you will notice that there are three separate bodies – one for each of the top, middle, and bottom faces.

9.   To connect them all, we use the **Stitch** tool. Click on **MODIFY**, then the **Stitch** tool.

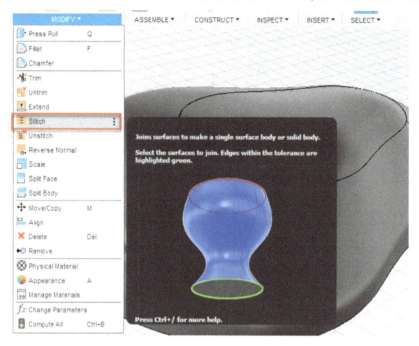

Figure 11.50 – The Stitch tool's location

10.   Select each body and then click **OK**. The three separate surfaces are now combined into one solid body.

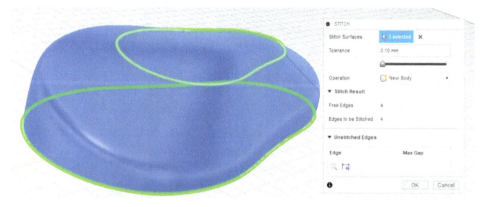

Figure 11.51 – The STITCH tool options floating window

We can now use all the body manipulation tools within the **CREATE** environment and take this body even further – cut more holes, create an empty body using the shell tool, and place inner parts such as a circuit board, and so on. I will leave it up to you how much further you want to take this model.

This was just one way to build a body using the **FORM** tool. We could have started with a cube and built from there, or we could have started with the **Surface** tool and built with it. Experimentation is the best way to learn, so look for items around your home and try to build them. Don't worry about trying to get them correctly built the first time. It's okay to make many mistakes since you are still learning. The most important thing is to keep building and creating.

Summary

The lessons we learned in this chapter were how to work within the **FORM** environment. As you can probably tell, there are many tools located within here that could take up a whole book on their own. We merely scratched the surface of the things that could possibly be created using the **FORM** environment. We learned how to create shapes such as a simple **Surface** box and **Form Box**, how to manipulate them with the **Edit Form** tool, and how to briefly work with **Surface** tools such as **Patch** and **Stitch**. All the tools looked at in this chapter are the basics of editing and creating 3D organic forms. Practice with these tools every day and you will soon become a master organic modeler.

In the next chapter, we will learn how to create a Tealight Scary Ghost using more **Surface** modeling tools.

12
Modeling a
Scary Tealight Ghost

In this chapter, we will explore the **Surface** tab and create a small tealight with a see-through ghost to fit over it. We will use the **Extrude**, **Revolve**, **Patch**, and **Thicken** tools to create both the ghost and the tealight. We will then finish off the model with a rendering with textures.

Figure 12.1 – A finished model of a tealight ghost

In this chapter, we will cover the following main topics:

- Using surfaces to create the tealight
- Creating the ghost with surface tools
- Rendering the ghost

Technical requirements

You can practice with the files provided or feel free to create your own for a more custom experience. The sample design for this chapter can be found at https://github.com/PacktPublishing/ Improving-CAD-Designs-with-Autodesk-Fusion-360/tree/main/Ch12.

Using surfaces to create the tealight

We will start out with a simple tealight candle for reference. We will create a component first to stick to Rule #1 (see *Chapter 3*) and then create the surface body, with a simple circle extrusion and a revolve for the candlelight. Let's get started:

1. Start a new design file, open the **Data** panel, and create a folder named Ch12 Tealight Ghost. Then, within the Chapter 12 folder, create a file and name it Tealight Ghost, and then set the units to millimeters.

2. Create a component named tealight, then create a sketch on the *XY* plane, and add a circle with a diameter of 37.5 mm.

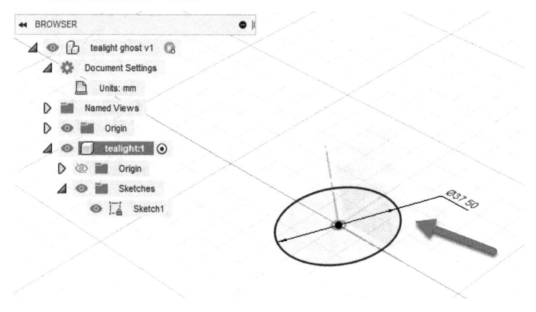

Figure 12.2 – Creating a tealight component and sketch

3. Click on the **SURFACE** tab, then the **CREATE** dropdown, and then **Extrude**.

4. Select the circle profile and set **Distance** to 12 mm. Note that the interior is empty. That is because surfaces have no mass, which means there is no weight to them. Think of them as a thin piece of tinfoil. Click **OK** to finish the command.

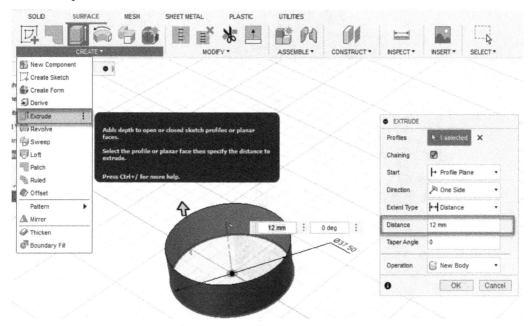

Figure 12.3 – Extruding a surface body

5. Select the **CREATE** drop-down arrow, then **Patch**, and then the bottom edge. Repeat the same for the top edge.

This fills the hole, but it does not mean that this cylinder is a solid model. It just places another surface where the hole is. The interior of the circle is still hollow. This is great to create reference bodies that generate a large file size and are quick to model.

6.   Click **OK** to finish the command.

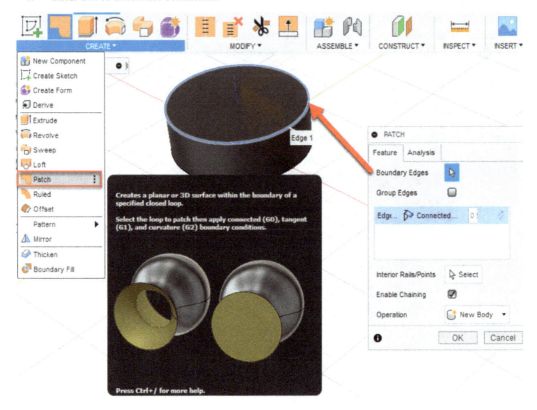

Figure 12.4 – The Patch tool selections

We will create the candle next.

7.   Select the **Create Sketch** tool and select the *YZ* plane. If it is hard to select the plane, hover the mouse over the area that you want to pick and hold down the left-click button for two seconds, and then a filter menu will appear. You can then easily pick the item from the drop-down list.

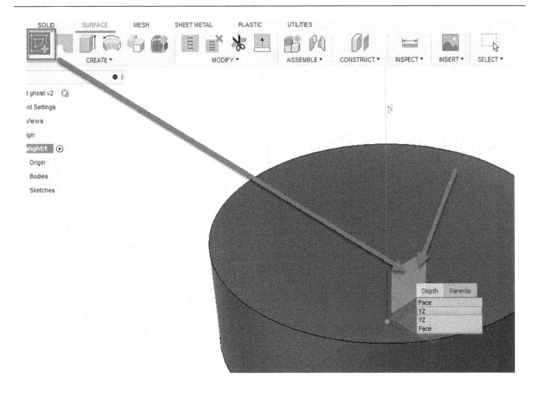

Figure 12.5 – Using the filter selection to pick a plane

8. Create a construction line at the center of the cylinder vertically and horizontally, and then use the **Fit Point Spline** tool to create a shape like the one in *Figure 12.6*.

9. Add the dimensions, as shown in *Figure 12.6*, and note that the spline is still blue, even though it is constrained.

    The reason for this is that the points have handles that can still shape the curvature of the spline. If you drag any of the handles, the points will not move due to the dimensions, but the shape will still change. You could attach constraints to them, such as horizontal or tangent, but in this instance, you don't have to.

10. Click on **FINISH SKETCH**.

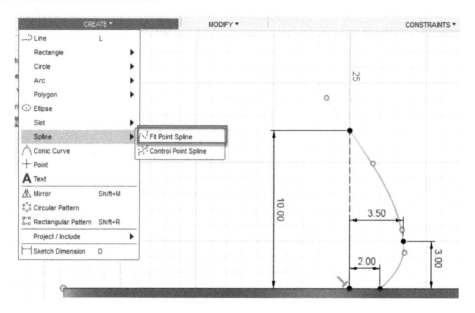

Figure 12.6 – The Fit Point Spline tool

11. Click on the **Revolve** tool, the spline for the profile, and then the vertical line for the axis. The candle is created from a single line, not a closed profile. If it was a closed profile, the candle would be a solid model. Click on **OK** to finish the command.

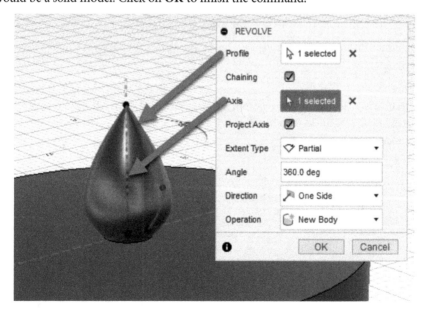

Figure 12.7 – The surface REVOLVE tool

Since the candle is a reference for our ghost character, we can leave it as it is now. We just needed the candle for its size. If we wanted to 3D-print this candle though, we would need to add thickness to the base and the candle, which would turn the candle into a solid model. Let's continue to create the ghost using more surface tools.

Creating the ghost

We will be using some more surface tools to create our ghost character. Feel free to go off-script if you want and create a ghost that you would like to use, using the same style as the one created here. Let's get started:

1. We need to create a new component by first activating the top level of the assembly. Hover the mouse over the tealight ghost, and select the dot to fill it in and make it active.

2. Now, create a new component and give it the name Ghost.

Figure 12.8 – Creating a new component

3. Click on the **Create Sketch** tool and then select the front face. Be sure that the **Ghost** component is active.

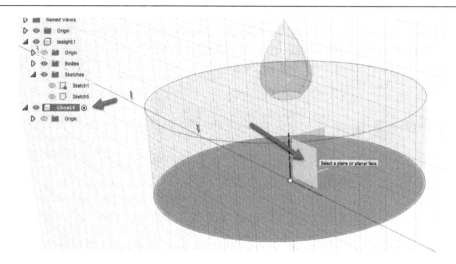

Figure 12.9 – Creating a new sketch on the front plane

4. Create another construction line from the midpoint vertically and horizontally, and then click on the **Fit Point Spline** tool once again, creating a shape that is similar to the one shown in *Figure 12.10*.

    Note that, this time, I selected the top horizontal fit point handle and added a horizontal constraint to keep it from rotating. This is to keep the top of the ghost's head from becoming an odd shape.

5. Click on **FINISH SKETCH** to complete the sketch.

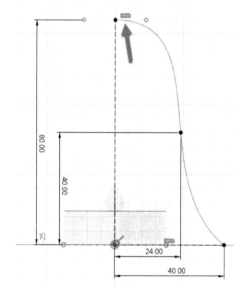

Figure 12.10 – The ghost sketch outline

6. To create the ghost body, go to the **CREATE** drop-down arrow, select **REVOLVE**, choose the outside curved spline, and then select the inner vertical construction line for the axis. Click **OK** to complete the command.

Figure 12.11 – Creating the ghost body with the surface REVOLVE tool

7. We will create the ghost's face now. Click on the **CREATE SKETCH** tool and choose the *XZ* plane. If you have trouble selecting the plane, click and hold the mouse for two seconds over the object you want to select, and the filter window will appear.

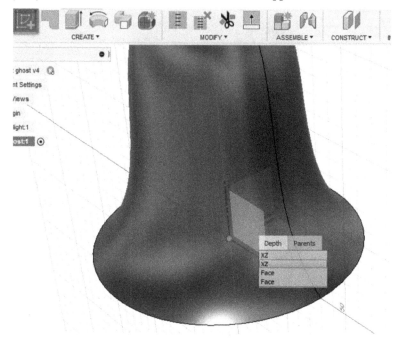

Figure 12.12 – Selecting the XZ sketch plane using the selection filter tool

8. For the face, create an ellipse for the eyes and a slot for the mouth, as shown in *Figure 12.13*.

9. To keep the eyes symmetric, create one ellipse on one side and add dimensions to it first.

10. Then, use the **MIRROR** tool from the **CREATE** panel and use the previous construction centerline (green arrow) as the axis to mirror the eye to the other side. Also, use this same construction centerline to place the slot as the mouth for the ghost.

11. Click on **FINISH SKETCH** to complete the sketch.

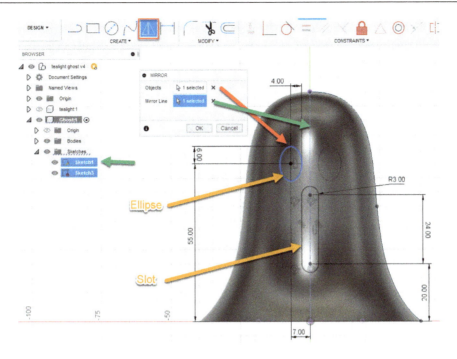

Figure 12.13 – Creating the ghost face using ellipses and slot tools

12. Click on **CREATE** and then **EXTRUDE** to extend the faces through the front face about 35 mm. When working with surfaces, you always overextend the pieces, as we will cut these back to create the cutouts using the **Trim** tool next. Click on **OK** to finish the command.

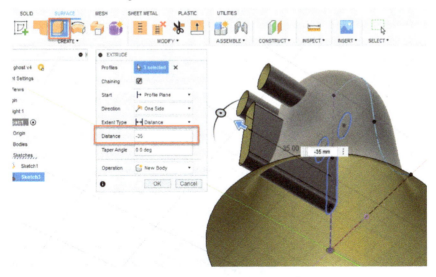

Figure 12.14 – Extruding the ghost face

13. Select the **Trim** tool within the **MODIFY** drop-down arrow and choose the ellipse (eye) body first, and then select the interior of the eye (the ghost body) next to remove that face. Repeat this process for the other eye and the mouth.

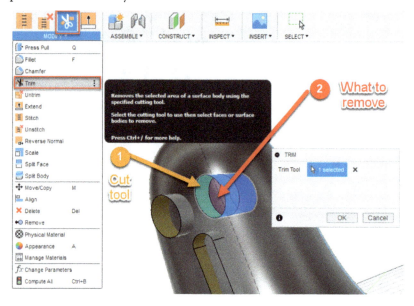

Figure 12.15 – The Trim tool selections

14. Start the **Trim** tool once again, and this time, select the ghost body first with the tool, and then select the eyes and mouth bodies to remove them.

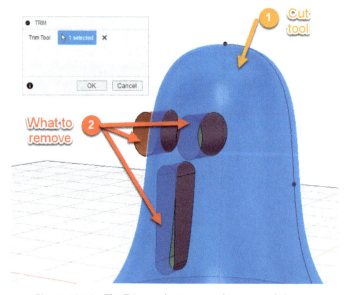

Figure 12.16 – The Trim tool removing the external shapes

15. Turn off the bodies for the eyes and mouth. We can keep these remaining interior shapes if we want to use them for indented eyes and a mouth.

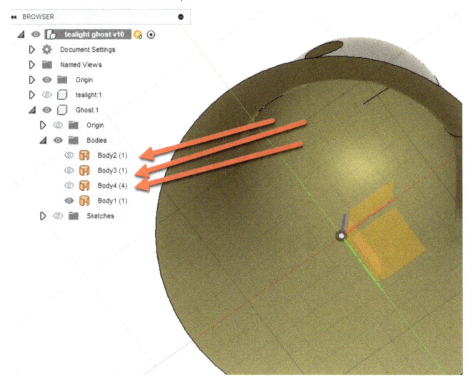

Figure 12.17 – Turn off the remaining eye and mouth trim shapes

16. Next, we will create arms for the ghost. Use the **Create Sketch** tool to create a sketch on the front plane.

17. Use the spline tool to create a shape like the one shown in *Figure 12.18*. It also may be helpful to turn off the body to draw the outline of the shape.

There are two things to note here. I purposefully did not connect the dots at 30 mm and 50 mm to the ghost body. This is because when working with surfaces, you need to overbuild and then cut back. The second thing is that when I create any of my splines, I use very few points. This helps to keep the shape simple and easier to manipulate with the fit point handles.

18. Click **FINISH SKETCH** to complete the sketch.

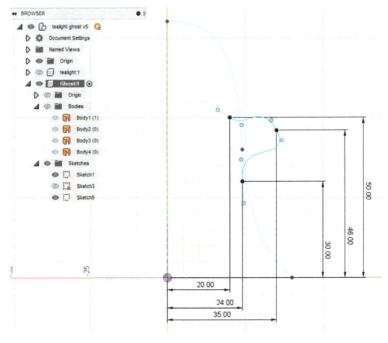

Figure 12.18 – Creating a hand outline using a spline on the front plane

19. Click on the **CREATE** dropdown, select **Extrude**, and then select the spline that was just created.

20. Set **Direction** to **Symmetric** and **Distance** to 17 mm, and then click **OK**.

Figure 12.19 – Extruding the spline hand

We could close off the front and back areas with the **Patch** tool, but we will create another sketch on the bottom plane and shape the hand a bit more.

21. Click on the **CREATE SKETCH** tool and select the bottom *XY* plane. Create a sketch like the one shown in *Figure 12.20* with the dimensions shown, using the **Fit Point Spline** tool.

I typically choose splines rather than arcs, as I can control the shape using handles and adjust curves much easier later on.

22. Click on **FINISH SKETCH**.

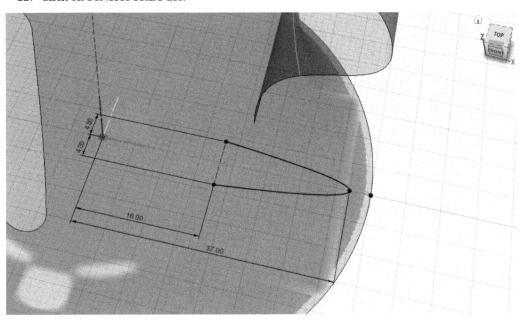

Figure 12.20 – A sketch on the bottom XY plane

23. Click on **Extrude** and select the spline. Set **Distance** to 56 mm and click **OK**.

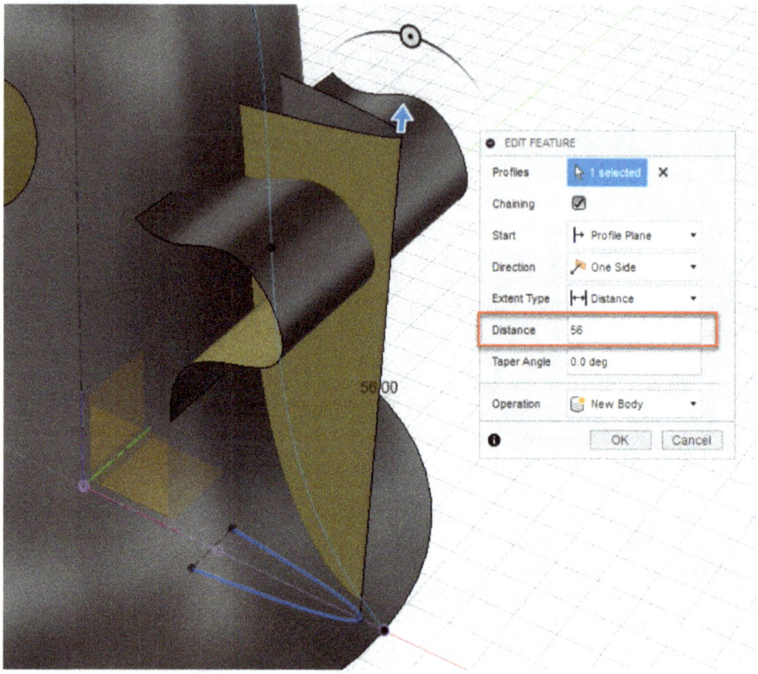

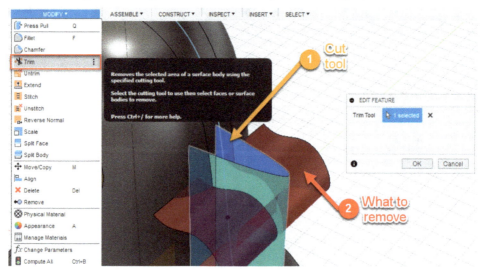

Figure 12.21 – Extruding the spline from the bottom XY plane

24. Click on the **Trim** tool, select the outer shape first, and then click on the arm shape to trim out the extra edges. This will patch the holes and shape the arm slightly.

Figure 12.22 – The Trim tool selection for the arm

25. Now, we will shape the arm more by reversing the selection and trimming the top of the arm Click on **Trim** again, select the inner arm as the **Trim Tool** object, then the outer extrusion (refer to *Figure 12.23*), and click **OK.**

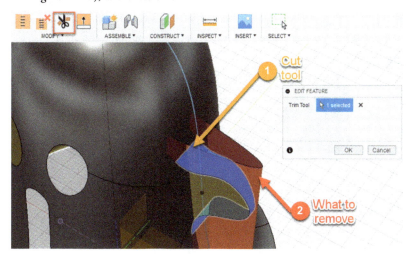

Figure 12.23 – The Trim tool to remove the outer extrusion

The arm looks pretty good on the outside. We still need to remove the inner edges though.

26. Orbit the model to view the bottom and click on the **Trim** tool again. This time, select the ghost body using the tool and then all the extra edges on the inside (see *Figure 12.24*).

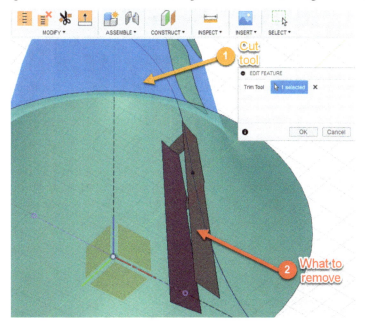

Figure 12.24 – Using the Trim tool to remove inner geometry

You may have noticed that the hand has two different colors. This can happen from time to time, as Fusion 360 may not know which side is supposed to face toward the screen. We will need to reverse the normal so that the gray part is on the outside and the orange is on the inside.

27. Go to the **MODIFY** drop-down arrow, select **REVERSE NORMAL**, pick the orange faces on the front and back of the hand, and click **OK**. All faces should be gray now.

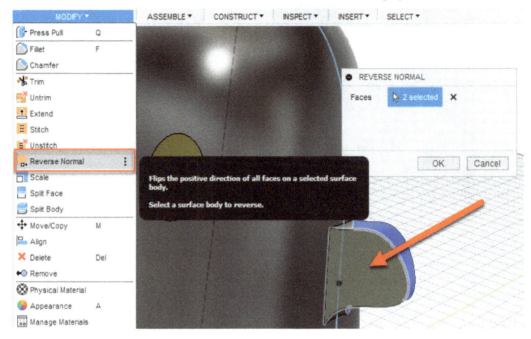

Figure 12.25 – The Reverse Normal tool selections

We now need to stitch together the hand so that all three faces are connected to make one surface body. If you open the dropdown in the Bodies folder within the **Ghost** component, you'll see a lot of surfaces. These are all the bodies we have created so far and some that have been trimmed away.

28. Go to **MODIFY** and then click on **STITCH**.

29. Click on the outer face first, then the inner, and then the other outer face, as shown in *Figure 12.26*. Click **OK** to complete the command. Now, the last few bodies have been combined into one.

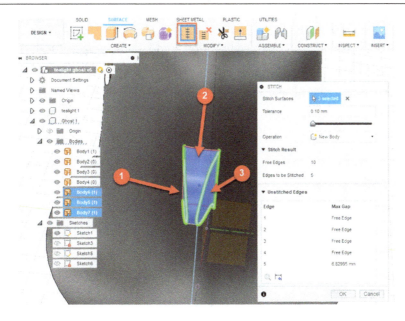

Figure 12.26 – The Stitch tool selection

Before we stitch the hand to the body, we need to mirror the hand to the other side.

30. Click on the **CREATE** drop-down arrow and select **MIRROR**. Choose the left hand for **Objects**, and for **Mirror Plane**, choose the *YZ* plane, and then click **OK**.

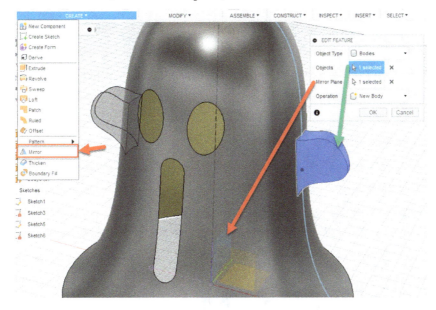

Figure 12.27 – The Mirror tool location and selection to create a right hand

Before we can stitch the hands to the body, we need to create a hole on the inside that the geometry can connect to; otherwise, the stitch will not work, so we will once again use the **Trim** tool.

31. Click on the **MODIFY** drop-down arrow, select the **Trim** tool, and then the left hand first (we can only do one hand at a time). Orbit the model to the bottom view and select the inner square on the ghost body. Click **OK**. Repeat the process for the right hand.

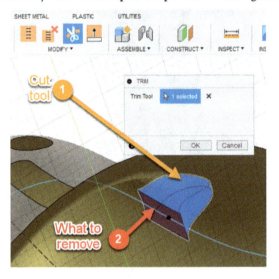

Figure 12.28 – Trimming the inner hole for the hands

32. We can finally connect the hands to the body. Select the **MODIFY** drop-down arrow, the **Stitch** tool, and then the hand, the body, and the other hand. Click **OK**.

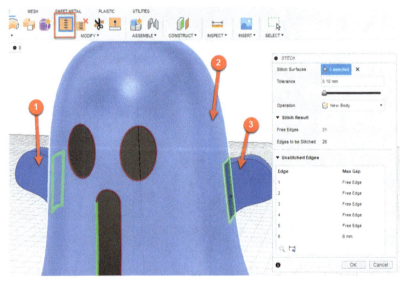

Figure 12.29 – Stitching the hands to the body

Now that everything is stitched together, we can add thickness to the body and turn it into a solid body.

33. Click on the **CREATE** drop-down arrow and select **Thicken**.

The **THICKEN** pop-up window will open, prompting you to select the faces. An interesting thing happens when you select the ghost body. Only the body is selected and not the hands, even though we just stitched it together. The reason for this is that the default selection for the **Thicken** tool is to select faces, not bodies, which is why only the body or hands is selected. We want to have everything selected at once, and to do this, we will open the Bodies folder within the browser and select the surface body that has everything stitched together (mine is **Body 11(7)**; your body number may be different to mine). This will highlight everything and let you thicken everything at once.

34. Click on the **Thickness** area. Note that if I provide a positive number, it will thicken toward the outside, and if I provide a negative number, it will thicken inward. Set a 1 mm positive thickness and click **OK**. A new solid body is added to the Bodies folder, which is our final result.

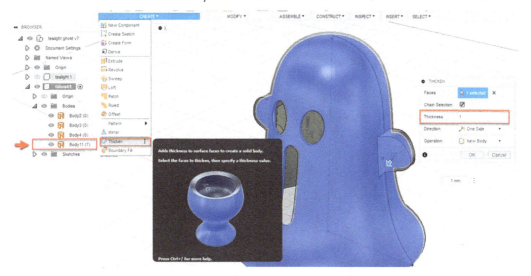

Figure 12.30 – Thickening the ghost

Let's add a fillet to the hands to smooth them out a bit, as they are currently rather sharp.

35. At the top of the screen, click on the **Solid** tab, located next to the **Surface** tab (since our bodies are now a solid model, we need to switch tabs). Then, click on the **MODIFY** panel and then the **Fillet** tool. Left-click on all six faces of the hands, add a 2 mm fillet, and click **OK**.

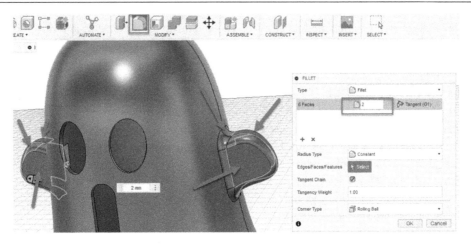

Figure 12.31 – The fillet tool location and selections

The final ghost model looks great! Let's bring it into the rendering environment and give it some texture and a scene befitting of a tealight ghost.

Rendering the ghost

The **RENDER** workspace is a great tool to showcase your models and show what they can look like in the real world with textures. We will use the same ghost model within the same workspace, so if you closed the ghost model file, open it back up. Click on the **DESIGN** workspace button at the top left and then select **RENDER**.

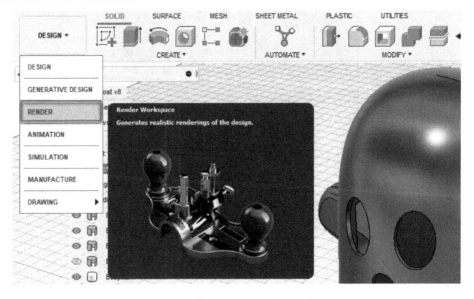

Figure 12.32 – The RENDER workspace location

Note that this environment has far fewer buttons than any other one we've worked on before. There is only the **SETUP** panel, the **IN-CANVAS RENDER** panel, the **RENDER** panel, and, within those drop-down arrows, not many other options.

Figure 12.33 – The RENDER environment overview

Our ghost seems to have a dark, shiny material applied to it. This is because all default materials in Fusion 360 start off as metal materials. There are two ways to change the material in Fusion 360, and we will look at these in the next section.

Changing the material

The two ways that you can change the material in Fusion 360 are as follows:

- **Physical Material**, which changes the physical properties of the material, such as from metal to wood. You would use this mainly for simulation purposes

- The second method is to change the material's appearance, which does not change the physical property, just the outer body material

We will work with both types to show the difference, so buckle up and let's get started:

1. Click on the **SETUP** drop-down arrow and then click on **Physical Material**.

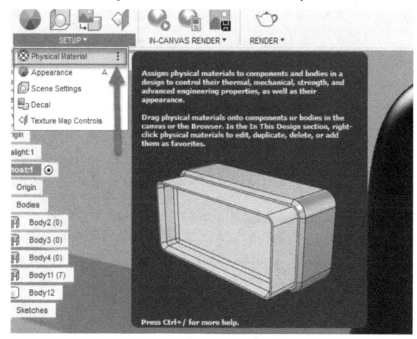

Figure 12.34 – The Physical Material tool location

The **PHYSICAL MATERIAL** pop-up window will appear, and on top of this flyout, you'll see an icon with an odd circle shape inside of it. This is the material that you are currently using in your design. If you double-click on it, a flyout will open, showing more options and providing info on the material's density. If you click on the **Advanced...** button, you can see more options on this material that you can view or edit. I typically trust these material options and don't make any changes to them, but if you need to make a change, you can select the icon, right-click, and copy to create another material to make changes on. Click on the **Cancel** button in the advanced options panel if you opened it, and then hit the **Done** button to bring you back to the **PHYSICAL MATERIAL** flyout.

Figure 12.35 – The PHYSICAL MATERIAL pop-up window

2. Click on the **Library** dropdown and choose **Fusion 360 Additive Material Library**.

    This library is for 3D-printing materials. If you were to 3D-print this ghost, these are the physical properties you would want to select for a particular material. Note that there are a bunch of typical plastics missing, such as PLA, but you can always edit and add your own if you want to run simulations.

3. For now, left-click and drag the top material onto the ghost body. The ghost will now change to a white color.

4.   Click on the **CLOSE** button at the bottom right to return to the workspace.

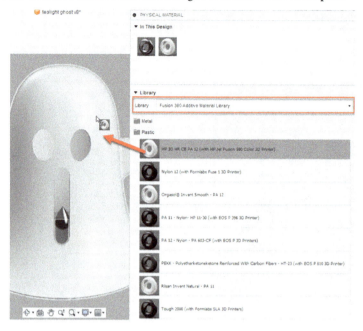

Figure 12.36 – The PHYSICAL MATERIAL Fusion 360 Additive Material Library

5.   Click on the **SETUP** dropdown and then **Appearance**. The **APPEARANCE** flyout window will open up.

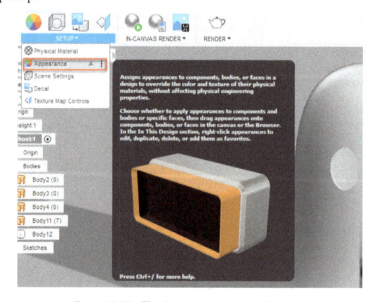

Figure 12.37 – The Appearance material location

Note that this window looks similar to the **PHYSICAL MATERIAL** window, but these materials are not representations of the physical material. To put it simply, these are like a different paint color on a car body.

6. Select the `Plastic` folder and then scroll down to the `Transparent` folder.

7. Left-click and drag the **Acrylic (Red)** color onto the ghost body. The ghost will now change to a clear dark red color.

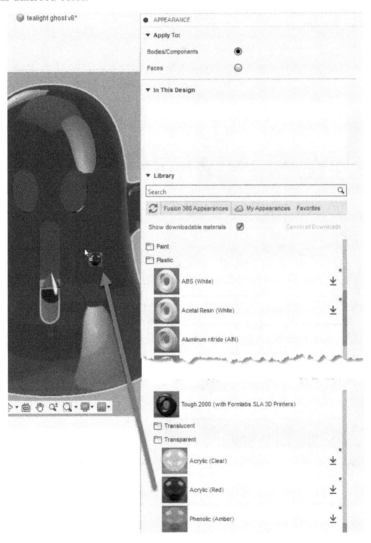

Figure 12.38 – The APPEARANCE flyout window

Let's lighten up the color a bit.

8. Go to the **In This Design** area, right-click on the **Acrylic (Red)** color icon that you just applied, and select **Edit**. A new will flyout appear with a color selector at the top.

9. Choose a new color if you want a different color, or you can adjust the numbers to match my selection, which is 238, 181, and 176. Be sure to change the name at the top to **CUSTOM** so that you know that the settings have changed from the original. The ghost is now a much brighter color than before.

10. Click on the **DONE** button on the **Edit** window and then select the **CLOSE** button on the **APPEARANCE** window.

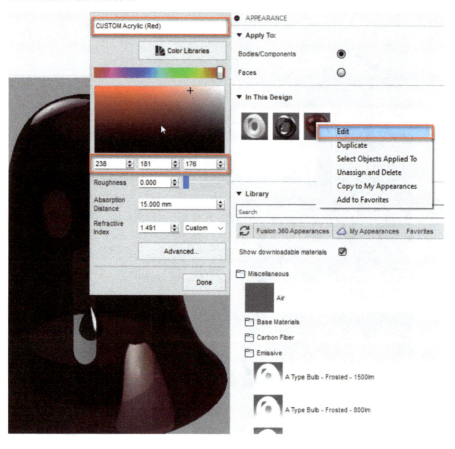

Figure 12.39 – Changing the color of a material

11. We can add materials to the tealight now. Hide the ghost body so that the tealight appears.

12. Go to **SETUP** and then click on **APPEARANCE**. In the **APPEARANCE** flyout, click on the `Miscellaneous` folder and then `Emissive`. Then, left-click on the **LED – SMD 3528 – 8lm (white)** material and drag it onto the candle.

A bright white color will appear, but this light can be a bit deceiving, as this bright color is only the material. It is not a true light. Fusion 360 does not have the ability to create separate lighting systems as most other 3D programs, such as Blender or Maya. We can just add this material to give the illusion that it is a bright light for the rendering.

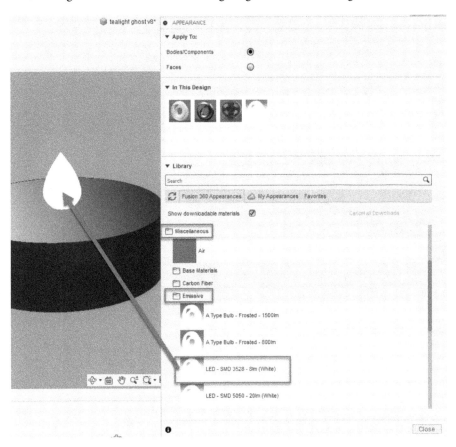

Figure 12.40 – Adding an Emissive material

13. Scroll down to the `Plastic` folder and drag out **Plastic – Glossy (Red)** and **Plastic – Glossy (Yellow)**. Drag them to the faces, as shown in *Figure 12.41*. Feel free to get creative with the colors and selections of your model if you wish.

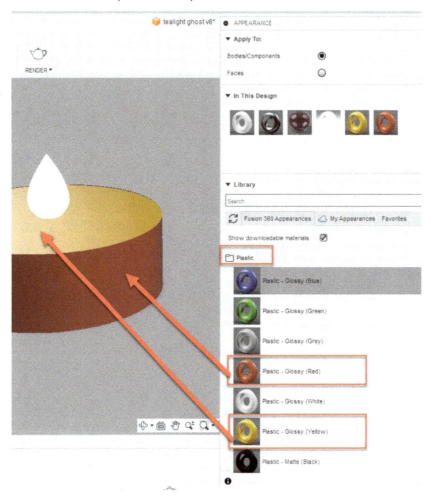

Figure 12.41 – Adding Plastic materials

The ghost model looks much better now, but let's change the background a bit in the next section.

Change the scene settings

The ghost model will look much better with a different background, as the plain gray doesn't look so great. Let's change it with the **Scene Settings** tool by following these steps:

1. Click on **SETUP** and then select **Scene Settings**. The **SCENE SETTINGS** flyout will appear.

2. Change the **Background** area from **Solid Color** to **Environment**.

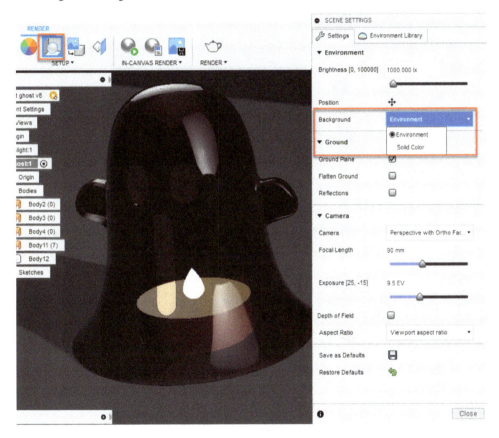

Figure 12.42 – The SCENE SETTINGS flyout

3. Select the **Environment Library** tab at the top and then pick a background from the library area below. I chose **Crossroads**, but note that I will need to click on the download arrow to apply it to my scene.

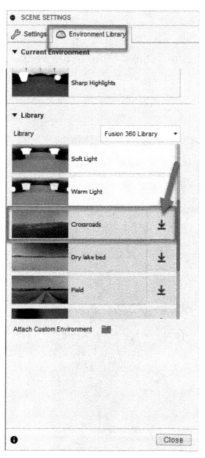

Figure 12.43 – Choosing an Environment background

Move your mouse into your scene, and middle-mouse scroll or use the zoom icons at the bottom of your screen; note that the background doesn't change when you zoom in or out of your model. Your model only changes its size.

4. Now, right-click and drag to orbit the model, and note that your model orbits as well as your background scene.

5. Click on the **Settings** tab and then the **Flatten Ground** checkbox. Your model now immediately connects to the ground plane of the image, and if you orbit and zoom now, your model adjusts to its environment.

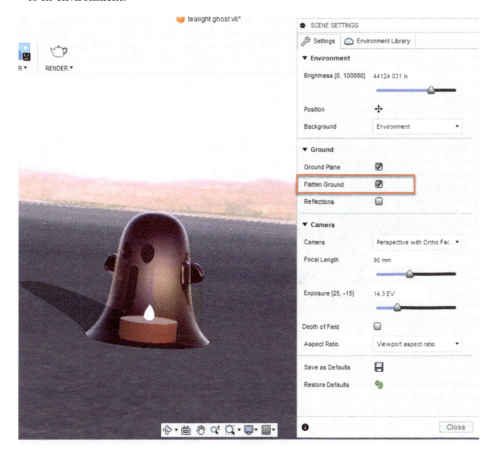

Figure 12.44 – Choosing an Environment background

6. Click on the **Brightness** slider and drag it to about **11000** to darken the scene a bit for our ghost. This will also make the candle pop a bit.

7. Click on the **Position** icon (the green box in *Figure 12.45*), and note that a slider appears at the bottom (the green box).

8. Click on the **Rotation** slider to adjust the background to a pleasing position in your scene. For me, the location **–22.0 deg** shows the road and background, which I'm happy with.

9.   Click on **Close** to keep these settings, or feel free to play and adjust more if you wish.

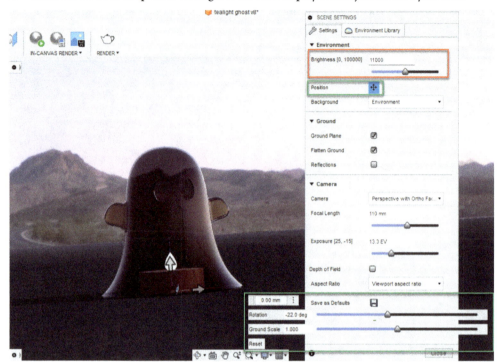

Figure 12.45 – Choosing an Environment background

Our scene looks much better now. Our final task is to render out an image so that we can send it to our friends or clients.

Render settings

There are two different render types in Fusion 360 – **IN-CANVAS RENDER** and **RENDER**. **IN-CANVAS RENDER** will quickly render your scene on your screen. So, follow along with me to render your scene:

1.   Click on the **IN-CANVAS RENDER** button, and your screen will change slightly; a timer bar will appear in the bottom-right corner of your screen.

      This is a live rendering. Do not touch your mouse, and let ithe render complete until it hits the **Excellent** bar.

2.   Now, rotate your model, and note that your screen will pixelate slightly. Then, restart and re-render the scene. This is a great way to make some visual changes while you render in real time.

Figure 12.46 – The IN-CANVAS RENDER screen

3. Click on the **Capture Image** button if you want to save the image to your Fusion 360 folder.

4. Click on the **Render Stop** button to go back to the render environment.

5. Click on the **RENDER** icon, which will look like a teapot now. The **RENDER SETTINGS** pop-up window will appear with a few different settings.

    Note that you could render in the cloud if you wanted to use cloud credits. These credits can be bought through your Autodesk account and let you send your render to a server, allowing you to continue working while your render is created elsewhere.

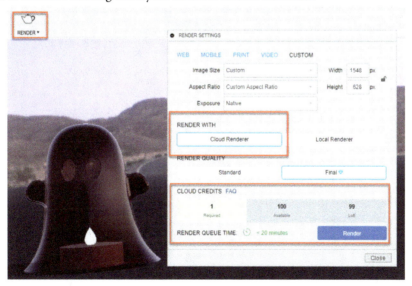

Figure 12.47 – The RENDER SETTINGS pop-up window

6. I typically do not have cloud credits and do all my rendering on my machine. Click on the **Local Renderer** button and the **CLOUD CREDITS** area will disappear, allowing you to render on your machine.

7. Click on the **PRINT** tab, choose the **4x6" 300PPI** option, and then click on **Render**.

Figure 12.48 – Changing to Local Renderer

A rendering of your scene will appear in the **RENDERING GALLERY** window at the bottom of your screen, with a bar indicating the time spent until the rendered scene is complete. You can create as many different renderings as you wish, and they will all be stored at the bottom of your screen. If you want to download any of your amazing renderings, click on an image and then the download button.

Figure 12.49 – RENDERING GALLERY at the bottom of the screen

Great job on completing your tealight ghost! I would love to see what you created by posting on any social media apps, which you can then hashtag-link to my name on my LinkedIn web page.

Summary

In this chapter, we learned how to create a tealight and a ghost with surfacing tools. We learned how to trim parts of surfaces by using other surfaces as trim bodies, and we learned how to stitch surfaces back together to form a new surface body. We then thickened the ghost, turning it into a solid model, and then added some textures to the ghost's body and tealight. Finally, we added a background and rendered the scene. Surfaces are a great way to create organic shapes with curvature, and unlike the **Form** environment, these surfaces can be parametric.

In the next chapter, we will use form modeling and solid modeling to create a cushioned chair.

13

Using Form and Solid Modeling to Create a Cushioned Chair

Being able to work in multiple environments is a great way to utilize all the tools in Fusion 360. We will explore how to create a chair shape in the Form environment and then bring it into the DESIGN environment to create a wooden chair frame within the form. We will discover some amazing abilities and find some shortfalls as well. Please be aware that this is a freeform modeling environment, which means that the models and dimensions may vary slightly from the models that you create. Try your best, and adjust your model and dimensions using the images and video as a guide. Remember that this is all just practice, and practice makes perfect.

In this chapter, we're going to cover the following main topics:

- Creating a chair in the Form environment
- Creating the inner frame
- Improving the design

Technical requirements

You can practice with the files provided or create your own for a more customized experience. The sample design for this chapter can be found at `https://github.com/PacktPublishing/Improving-CAD-Designs-with-Autodesk-Fusion-360/tree/main/Ch13`.

Creating a chair in the form environment

In this chapter, we will create a floor gaming chair such as the one shown in *Figure 13.1*. The reason for creating the chair in the Form environment is that for a shape like this, it is much easier to create it using forms rather than trying to build it using a parametric model. For the interior framework in the next area we will use parametric modeling, since the interior framework is a more rigid structure, which parametric modeling is ideal for.

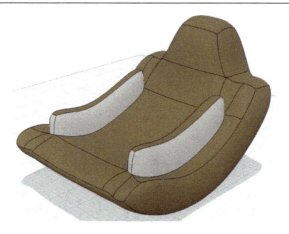

Figure 13.1 – The floor gaming chair final image

Let's start by opening the **Data** panel, creating a new folder named Ch13 Gaming Chair in the PACKT pub project folder, and saving the file as Gaming Chair. Set the units of the drawing to inches:

1. Go to the **CREATE** panel and click on the **Form** button. Then, go to the **CREATE** drop-down arrow and click on the YZ plane.

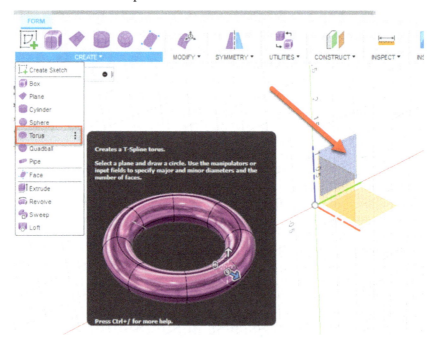

Figure 13.2 – The Torus tool location and plane selection

2.   Create the torus with the following settings:

- **Diameter 1** is 36 (this is the overall height of the chair)

- **Diameter 1 Faces** is **10** (these are the faces around the entire outside)

- **Diameter 2** is 12 (this is the inner circle size)

- **Diameter 2 Faces** is **6** (this is the number of faces for one tube section)

Click **OK** to finish the command.

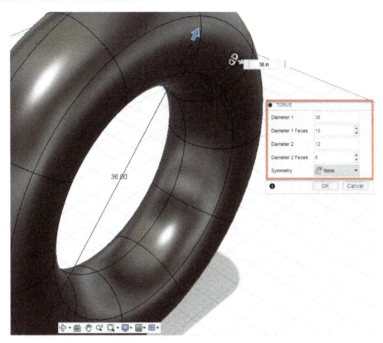

Figure 13.3 – The torus size settings

3.   Click on the ViewCube and select the right side to orient the screen to see the side of the torus. Select the top half and a small bit of the bottom. Be sure that **Select Through** is checked under **Selection Filters** in the **SELECT** drop-down arrow. This will make sure that both sides of the model are selected. Then, hit the *Delete* keyboard key to remove that half.

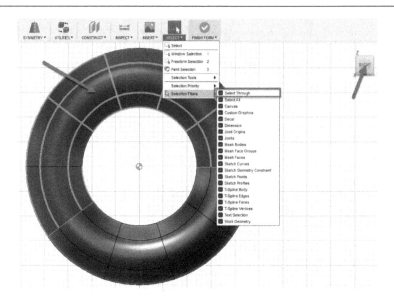

Figure 13.4 – The torus selection to remove

4. Orbit the view, then go to the **SYMMETRY** drop-down arrow, and select **Mirror - Internal**. Pick two faces on either side of an edge and a green line will show up. Now, when we select something on one side of the green line, it will also select something on the other. This is a great way to keep your models symmetric. Click **OK** to complete the command.

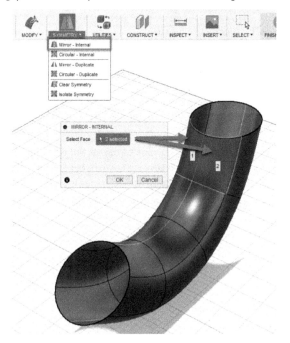

Figure 13.5 – The Mirror - Internal tool location and selection

5. We need to make the chair wider now. Click on the **FRONT** ViewCube and then select one-half of the torus. Go to the **MODIFY** drop-down arrow and select **Edit Form**. Using the directional arrow, drag the selected side to the right by 6 inches.

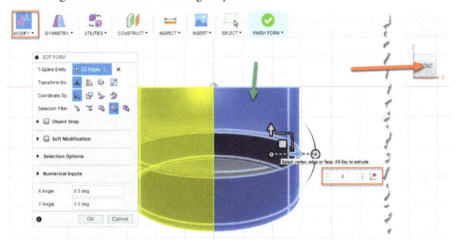

Figure 13.6 – Widening the torus chair with the Edit Form tool

6. We are now going to add some extra geometry that will let us shape the chair edges. Click on the **MODIFY** drop-down arrow and click on the **Insert Edge** tool. Double-click on any of the front center edge lines to select them all; any line will work, as double-clicking any edge will select multiple ones. It will select the entire ring, but since there is a hole at the top and bottom, double left-clicking will only select the front edge line.

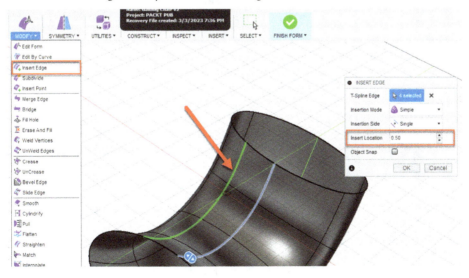

Figure 13.7 – Insert Edge on the front center line

7. Repeat the same process for the rear edge line using the **Insert Edge** tool.

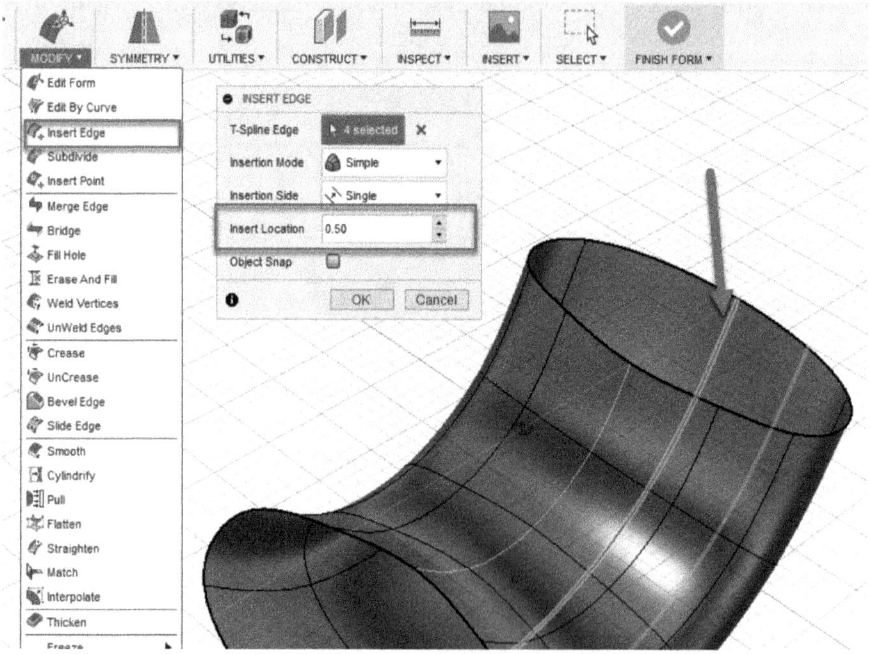

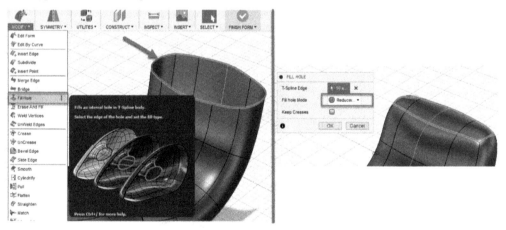

Figure 13.8 – Insert Edge on the back center line

8. Double-click on the top ring and note that this time, it selects the entire ring. Now, click on the **MODIFY** drop-down arrow and then **Fill Hole**. The **FILL HOLE** pop-up window will appear. Select the drop-down arrow under **Fill Hole** mode, select **Reduced Star**, and click **OK**.

Figure 13.9 – The Fill Hole tool

9. Repeat the same process for the bottom hole with the same **Fill Hole** mode.

Figure 13.10 – Fill Hole for the bottom hole

Shape the chair with the Edit Form tool

Now that we have the basic shape down, we can start to shape it further to make it look like a chair. Let's add the headrest and sides now using the **Edit Form** tool:

1. If you click on the **RIGHT** side of the ViewCube, you will see that we have what looks like a fat bean shape. We need to reduce that shape down to a thinner shape. Double-click on the center line to select the entire ring shape, and then drag down the top center scale bar. This will only scale along the Z axis.

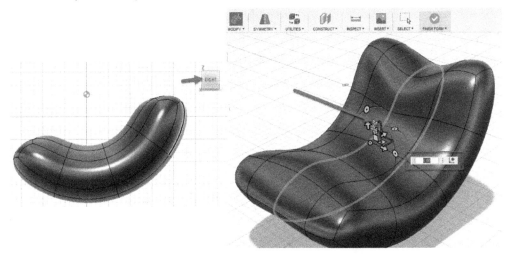

Figure 13.11 – Thinning out the center using the Edit Form tool

2. Without closing the **Edit Form** tool, double-click on the line to the right to select that ring, and select the same scale line to scale down along the *Z* axis by about . 6 units.

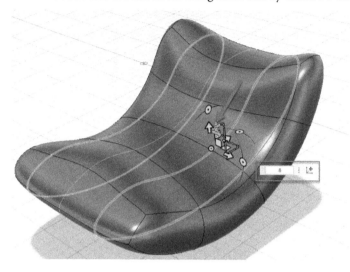

Figure 13.12 – Thinning the shape further using Edit Form

3. The front area still has a little bit of a bump. Let's grab that face and move it down a bit. Click on the face closest to the bottom, drag on the *Z* directional arrow, and drag it down by about −3 . 5 units.

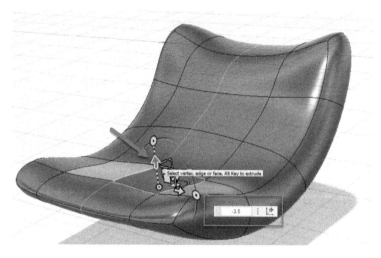

Figure 13.13 – Moving a face down in the Z direction

4. Let's add the headrest next. Click on the top face and hold down *Alt* + drag in the positive *Z* direction (holding down *Alt* + drag with any **Edit Form** modification tool will add geometry). However, note that it creates two separate headrests instead of one (see *Figure 13.14*). We do not want this though, unless you're creating a gamer chair built for two people. Hit **Cancel** to remove that selection. If it remains after hitting Cancel, you can use the keyboard shortcut for **Undo**, which is *Ctrl + Z*.

Figure 13.14 – Holding down Alt + drag to add geometry while Mirror - Internal is on

5. Turn off symmetry by selecting the model first, going to **SYMMETRY**, and then clicking on **Clear Symmetry**.

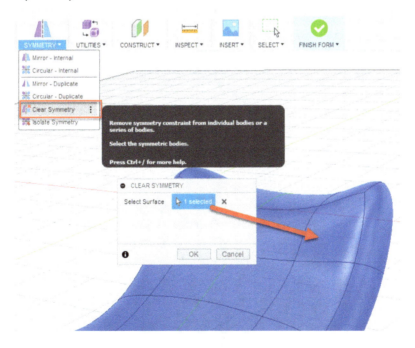

Figure 13.15 – Removing symmetry with Clear Symmetry

6. Now that the symmetry has been removed, we can add a single headrest by selecting both top faces, holding down *Shift*, and selecting each one. Then, select **Edit Form**, hold down *Alt* + drag, and click on the *Z* directional arrow in the positive direction about 10 units, which will create a single headrest.

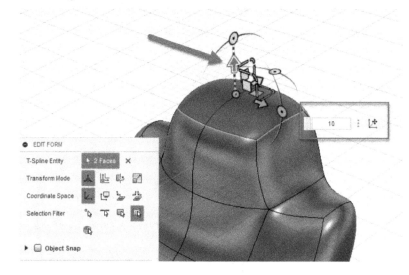

Figure 13.16 – Creating a single headrest without symmetry turned on

7. With the same faces still selected, click on the **Scale all** dot, and drag inward about .7 units to shrink the headrest a bit. Click **OK** to finish.

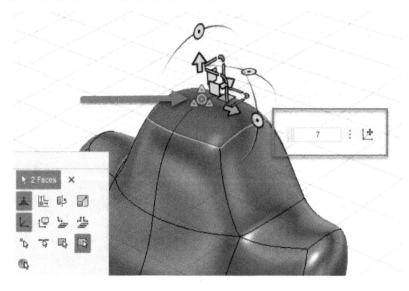

Figure 13.17 – The Scale all directions dot location

8. Now, let's add the armrests, but to make our lives easier, let's turn symmetry back on so that our changes happen on both sides of the model. Go to **SYMMETRY**, then **Mirror - Internal**, and select two faces on opposite sides of the center line. Click **OK**.

Figure 13.18 – The Mirror - Internal symmetry selection

9. We need to add geometry to the sides to create our armrests, but we need to keep our geometry clean-looking by making sure the new edges go all the way around our model. Double-click on the edge closest to the center line and then click on **Insert Edge**. However, note that our selection does not pick the top edge. If I were to select that extra line, it wouldn't be added to the current selection. This is OK; we can add it another way. Click **OK** to insert the edge that we currently have selected.

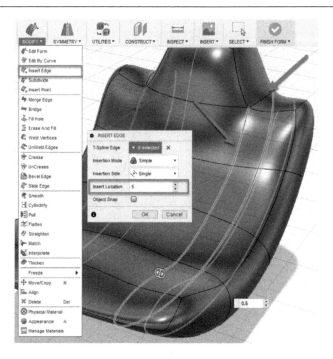

Figure 13.19 – The Insert Edge selection

10. Reselect the **Insert Edge** tool and select the edge at the top. Set **Insert Location** to −1, and note that it places it close to the other edge in the preview. Click **OK** to place that edge.

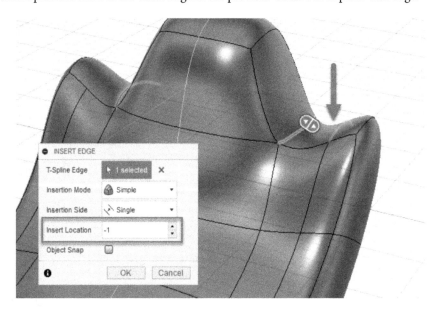

Figure 13.20 – The INSERT EDGE selection for the top edge

11. The edge is close to where it should go, but it is not connected to the other edge that was created earlier. We need to connect those two edges together to have clean geometry – that is, we do not want to have random lines all over our model; instead, we want them all to connect to each other. Select **MODIFY**, and then **Weld Vertices**. Note that your model will now show all the vertices.

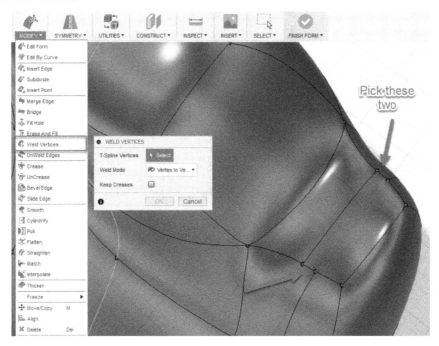

Figure 13.21 – The Weld Vertices tool location

12. Select the back two vertices and note that once those two are selected, your model will go into Box Display Mode. Your model may do this from time to time, and what this means when it happens is that something may have become disconnected or misaligned, so Fusion 360 will change to Box Display Mode so that you can see the error. Do a window selection on the bottom two dots (they are currently on top of each other,) and your model will go back into Smooth Mode. Click **OK** to finish the **Weld Vertices** command.

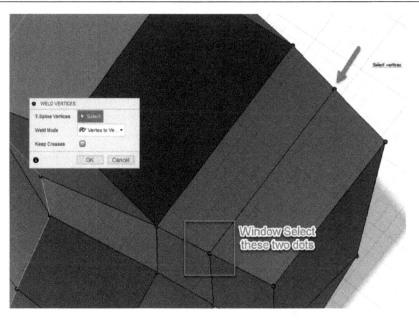

Figure 13.22 – Box mode during the weld selection

13. To add the armrests, we will use **Edit Form** and *Alt* to add more geometry. Click on the **Edit Form** tool, and hold down *Shift* to select multiple faces.

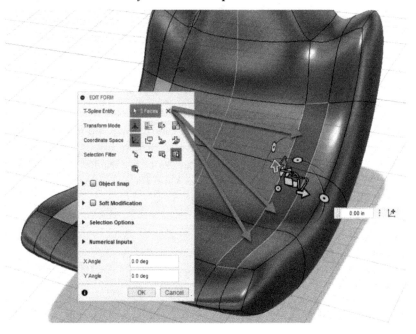

Figure 13.23 – Selecting multiple faces using Shift

14. Hold down *Alt*, and drag the *Z* direction arrow about 5 units. Click **OK** to finish the command.

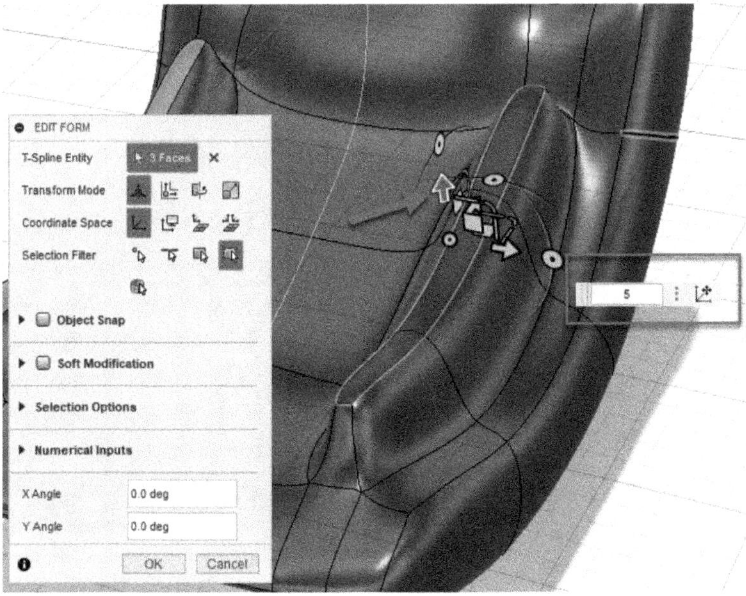

Figure 13.24 – Edit Form with Alt and dragging the Z directional arrow

15. I would like to add a hard edge around the outside of the armrest. We can use the **Crease** tool to accomplish this. Hold down *Shift* and select the edges around the armrest; don't forget to select the top and bottom edges as well.

Figure 13.25 – Edge selection for the Crease tool

16. Go to the **MODIFY** drop-down arrow and select **Crease**. Your model will turn yellow with dark yellow dots. Click **OK** to accept the selection.

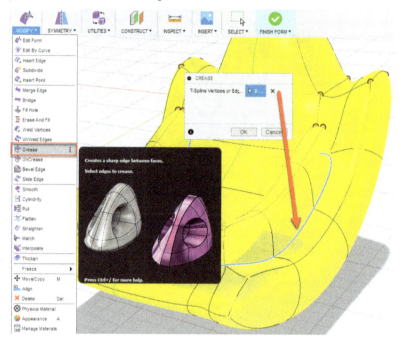

Figure 13.26 – The Crease tool location and selection

17. The model looks great. Continue to go around and adjust the model as you wish; be sure to save before you continue to give your model a point of reference to go back to if something goes wrong. I lowered my shoulders a little bit and adjusted some points to get the model shown in the following figure. Once you are happy, click on the **FINISH FORM** green checkmark.

Figure 13.27 – The finished form model

We will now add some materials to the faces of the model. This will give us a better idea of what the finished piece will look like.

Adding a material to faces

We are now out of the Form environment and back in the Create environment. You are not able to add materials within the Form environment, but since we are no longer modeling in that space, we can finish the form by going back to the Create environment:

1. Go to the **MODIFY** drop-down arrow and then **Physical Material**. We will change the material from a metal one to a fabric one. Click on the `Fabric` folder, and then drag and drop onto the model the **Linen, Beige** material. This changes the material properties of the model and the way that it looks. Click **Close** to finish using the **Physical Material** tool.

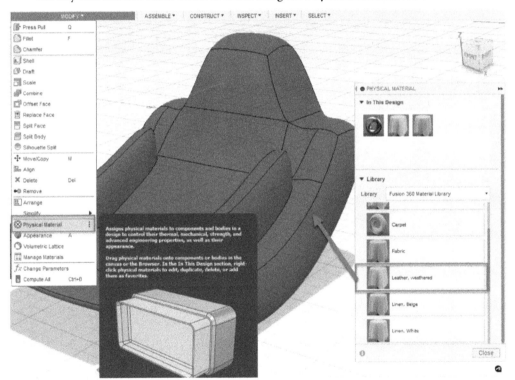

Figure 13.28 – The Physical Material tool location and selection

2. Click on the **MODIFY** drop-down arrow again, and this time, select the **Appearance** tool. At the top of the **APPEARANCE** palette window, select the radio button next to **Faces**. Then, in the **Library** area, select the Leather and Cloth folder, choose the Leather subfolder, and then click on **Leather - Matte (White)**. You may have to click the download arrow to use this material. Drag and drop this material onto the faces around the armrests. Click **Close** when you have finished your selections.

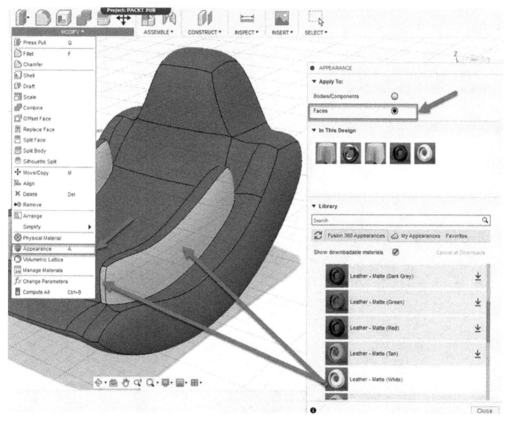

Figure 13.29 – The Appearance tool location and face selections

You can choose as many faces as you want to create the custom fabric chair of your dreams. Next, we will start to create an inner wooden frame.

Creating the inner frame

Let me start off by saying that I'm not a woodworker. I've worked in a shop environment and picked up a few tricks of the trade, but I am not by any means a full-on craftsman. So, the interior of this chair is my best guess of how I would make this chair. If you want to build a chair like this, please reach out to a craftsperson who I'm sure will help sketch out some great ideas:

1. Since we will add an internal framework, we can use the **Section Analysis** tool to view the interior of the model. Click on the **INSPECT** drop-down arrow and then click on **Section Analysis**. The **SECTION ANALYSIS** pop-up window will open, asking you to select a cut plane. Click on the *YZ* plane, and a preview of the section will open. Note that there are lines within the model. These are called hatch lines, which will change depending on the material used and represent the inside of the material. Since the lines go all the way through, this means that the object is a solid and not a hollow surface model. We need to hollow it out. Click **OK** to accept and close the **Section Analysis** tool.

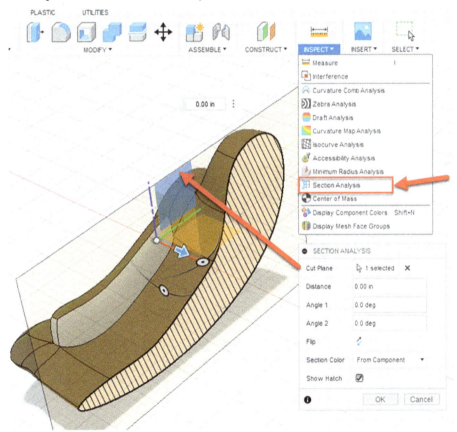

Figure 13.30 – The Section Analysis tool location

2. To create a void inside the model, we need to use the **Shell** tool. Turn off the **Section Analysis** tool by clicking on the eyeball icon in the browser next to the `Analysis` folder name. Go to the **MODIFY** drop-down arrow and then click on **Shell**. The **SHELL** pop-up window will appear. Select the model, change the direction from **Inside** to **Outside**, and then change the thickness to 1/8 of an inch. As long as the model has no issues, such as overlapping edges or pinch points, when you click **OK**, the model will grow by 1/8 of an inch on the outside.

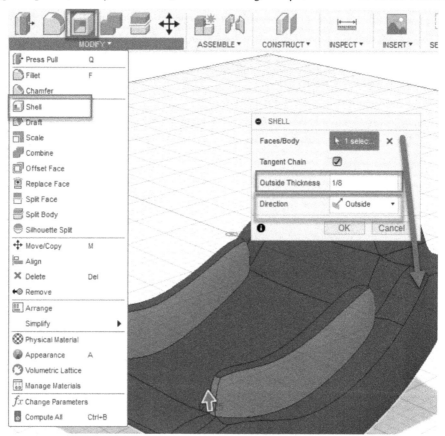

Figure 13.31 – The Shell tool location and selections

3. Turn on the **Analysis** eyeball icon now, and note that the **Shell** tool has worked – there is a 1/8 of an inch thin section all around the outer edge.

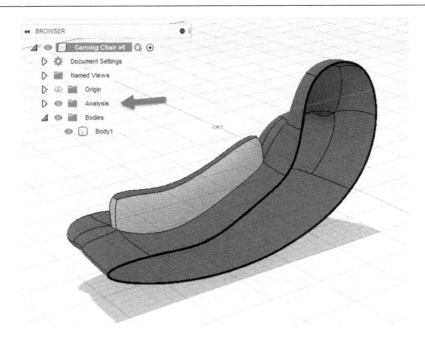

Figure 13.32 – Section Analysis turned on with the Shell tool

4. Now, we will create one of the support arms and then mirror it over to the other side. Open the `Analysis` folder and double-click on `Section1`. This will open the **EDIT** pop-up window. Drag the section arrow back to about −12.4 units. Click **OK**.

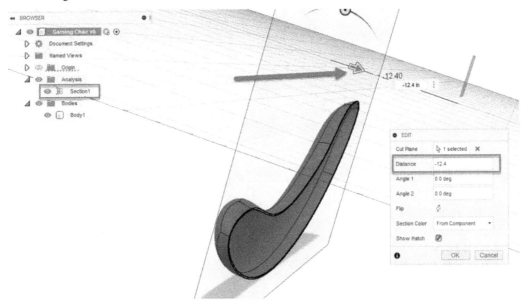

Figure 13.33 – Adjusting the Section Analysis plane distance

5. Click on the **Offset Plane** tool within the **CONSTRUCT** panel, and select the *YZ* plane. Set the distance to -12.4 in. Click **OK**.

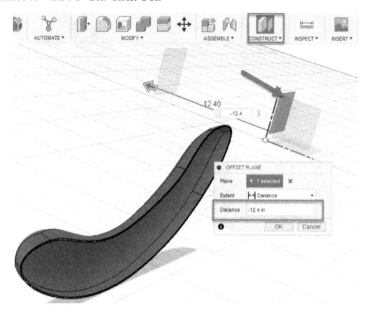

Figure 13.34 – The Offset Plane tool settings

6. Click on the **Create Sketch** tool and select the new plane that was just added.

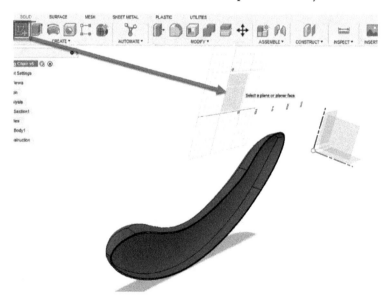

Figure 13.35 – The Create Sketch tool plane selection

7. Create a shape like the one shown in *Figure 13.36* using two Fit Point Splines and two Tangent Arcs. Note that I didn't use too many points while creating the Fit Point Splines. It was just a start and an endpoint spline to which I then adjusted the fit point handles to create the curves. Click **FINISH SKETCH** when complete.

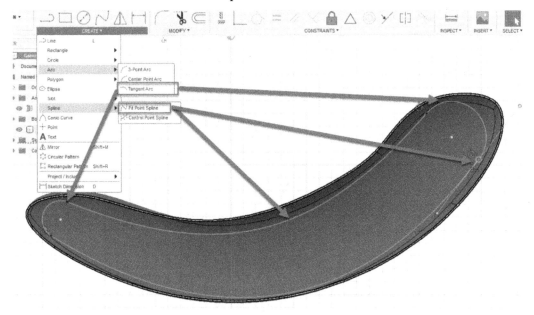

Figure 13.36 – Creating the roller arm support sketch

8. Select the **Extrude** tool within the **CREATE** panel, and select the sketch that was just created. Change the direction to **Symmetric** so that it extrudes on both sides of the sketch profile, and set the measurement to **Whole Length** and the distance to 1/4 of an inch. The extrusion goes through **Section Analysis** since we moved the plane. We can move that back now to see our extrusion.

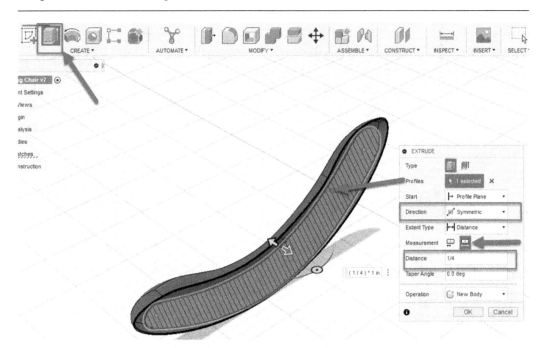

Figure 13.37 – The Extrude tool settings

9.   Open the Analysis folder and double-click on Section1. Note that the previous distance is gone, as it now starts from 0. If you don't remember the last distance that you used, you can hover the mouse over the right end of the distance name and left-click on the three dots to bring up a window of previous distances. Select **−12.4 in**, remove the minus sign to change it to **12.4 in**, and click **OK**. Now, you can see the halfway point of your seat and the extruded leg support.

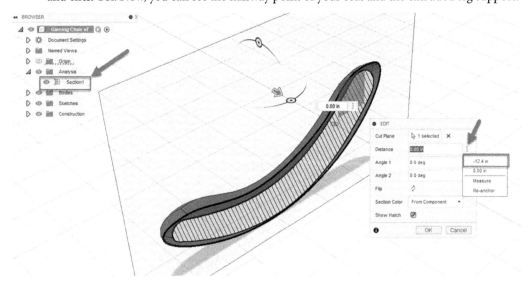

Figure 13.38 – The Section Analysis tool distance settings

10. We will mirror the analysis to the opposite side now. First, we need to turn off the analysis by clicking on the eyeball next to the folder name `Analysis`, which will turn off all analyses within that folder. We then need to turn off the chair model as well. Open the `Bodies` folder and click on the `Body1` eyeball (your name may be different) to turn it off for now.

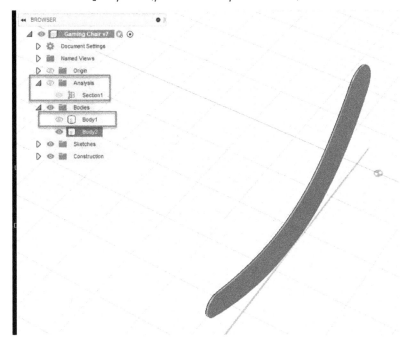

Figure 13.39 – Turn off Analysis and Body1

11. Before we mirror though, there is one important step that we forgot to add – create a component for these bodies! That's OK; we can still create components even if we forgot to do so at the beginning. Right-click on the `Body1` name and select **Create Components from Bodies**. This will create a new folder; Fusion 360 will name it `Component1:1`, and then Fusion 360 will place that `Body1` in that component automatically.

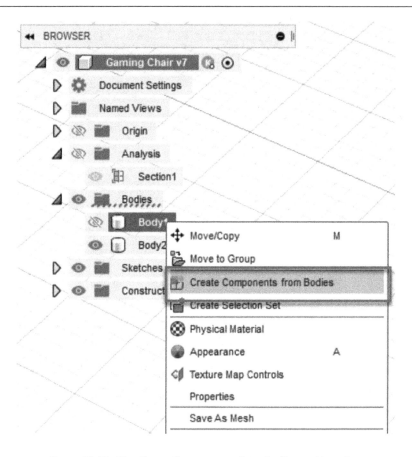

Figure 13.40 – The Create Components from Bodies tool location

12. Another way to do this is by manipulating the Timeline. Drag the Timeline slider back to before we created the sketch. Now, right-click on Gaming Chair at the top of **BROWSER**, and then select **New Component**. This will create a new component at a point in time before the sketch was created. Name the new component Chair Support and click **OK**.

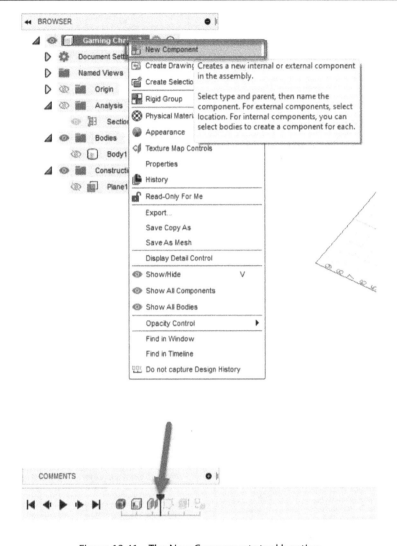

Figure 13.41 – The New Components tool location

13. This will create a new component in your browser, which automatically becomes the active component, and you will be placed in it to start drawing. However, there is nothing in your Timeline now. This is not a problem though; go back up to the top of **BROWSER** and activate the top-level assembly black radio button (*step 1*). Once that is done, the Timeline will reappear, showing everything. Drag the slider above the sketch so that it now appears in your browser (*step 2*). This reveals the Sketches folder. Now, drag Sketch3 (the name may vary) down into the Chair Support folder. This will move the sketch into that component now. Drag the slider all the way to the right now, and note that the body automatically moves into the Chair Support component, since that is where the sketch is located now. Finally, rename Component1:1 to Chair Material.

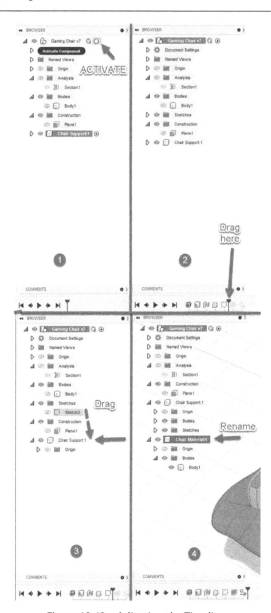

Figure 13.42 – Adjusting the Timeline

14. Activate the Chair Support component, and note that we can now see inside the chair. This is a great benefit of creating components, as we can now concentrate on the interior without having to worry about the **Section Analysis** tool. We can mirror the interior rail now. Click on the **CREATE** drop-down arrow and select **MIRROR**. Select the rail body and select the YZ mirror plane. Click **OK** to complete the command.

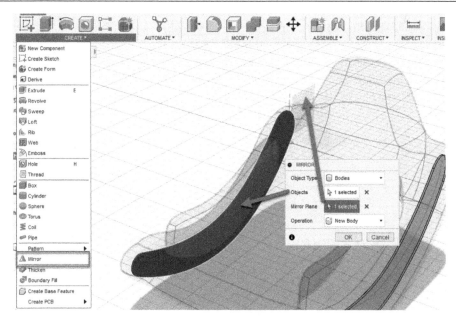

Figure 13.43 – The Mirror tool location and options

15. We will add some supports to the rails to give the chair some strength. Click on **Create Sketch** and select the face of the rail on the left.

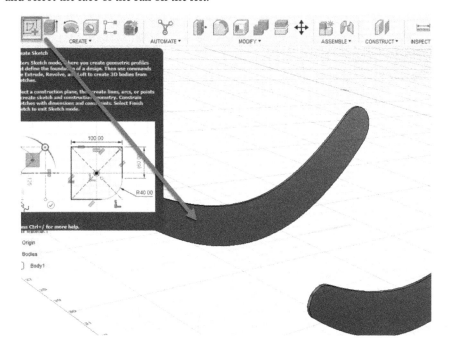

Figure 13.44 – The Create Sketch face location

16. Create a sketch like the one shown in *Figure 13.45*. I used the line tool to create the 2x4 box shape (the actual size of 2x4 is closer to something like 3.5 inches x 1.5 inches). The end dimensions are connected to the center arcs at the ends. The corners of the 2x4 box are all parallel and the sides are equal constraints. The top and bottom of the 2x4 box are also parallel constraints. Click on **FINISH SKETCH** when you are done.

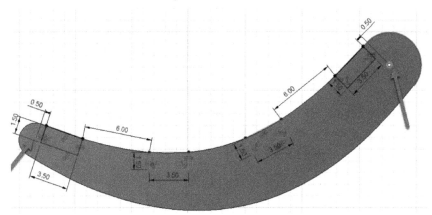

Figure 13.45 – The sketch drawing for the support cutouts

17. Choose **Extrude** from the **CREATE** panel, and select the sketch profiles that we just created. Set the extent type to **Object**, and select the opposite face of the body that the sketch was created on. This way, if that thickness needs to change, the extruded cut will also change size as well. Set the operation to **Cut** and click **OK**.

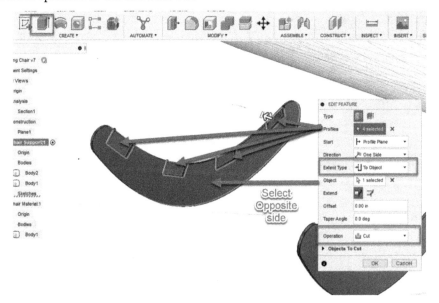

Figure 13.46 – Extruded cuts using sketch profiles

18. The cuts have been added to one side of the rocker legs, but they were not added to the other side. We can fix that by manipulating the Timeline at the bottom of the screen. Click and drag on the **Mirror** operation in the Timeline and drag it to the far right. Now, the cuts have been added to both sides, as the **Mirror** operation has been moved to after we added the cuts instead of before.

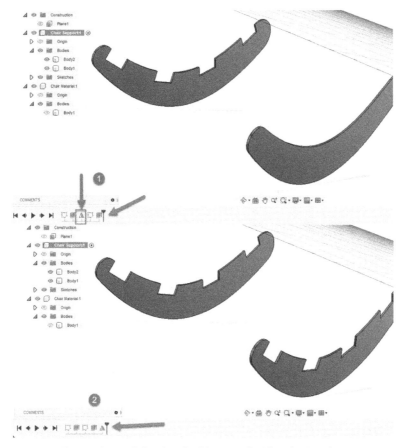

Figure 13.47 – Adjusting the Mirror tool within the Timeline

19. To create the rails, we can use the same sketch that we just created. Go to **BROWSER** and turn on the eyeball for the sketch that contains the cutouts (mine is Sketch1). Click on the **Extrude** tool and select the sketch profiles. Set the extent type to **To Object**, and select the far face of the opposite rocker leg. Set the operation to **New Body** and click **OK**.

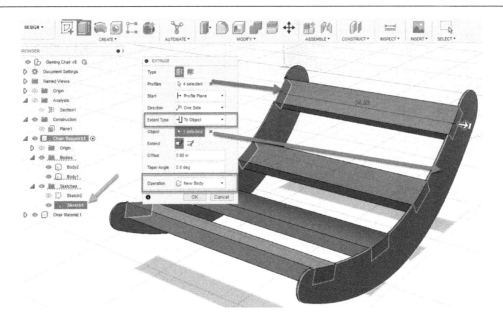

Figure 13.48 – Extruded supports from existing sketch profiles

The chair has a nice frame to support a person's back. Now, we will create the head and arm supports.

Head and arm support

For the head support, we will create another sketch that attaches to the lower frame:

1. Be sure you are still active within the Chair Support component. Click on the **Create Sketch** tool and select the *YZ* plane.

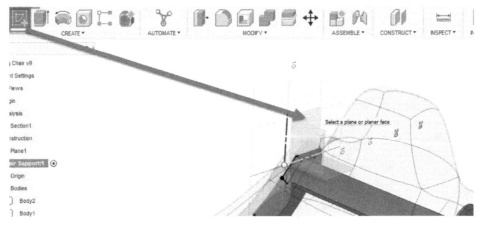

Figure 13.49 – The sketch profile selection on the YZ plane

2. Create a sketch like the one shown in *Figure 13.50*. Remember that you do not have to create this line by line. Create the basic shape first, and then add the constraints. Then, add the dimensions. You can use my drawing models for reference if you have trouble. Click **FINISH SKETCH** when you are done.

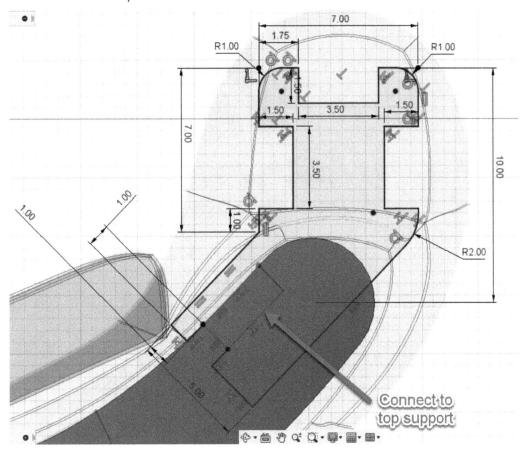

Figure 13.50 – The head support sketch profiles

3. Click on **Extrude** within the **CREATE** panel and select the sketch profile we just created. Set **Start** to **Offset**. This will allow us to create our 3D model away from the sketch plane instead of right on it. Set **Offset** to 4 in. Set the distance to 1/4, which is the thickness of the wood. Click **OK** to complete the command.

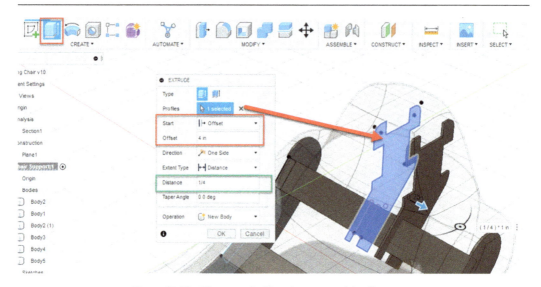

Figure 13.51 – The extruded head support with offset start

4. Mirror this part to the opposite side using the *YZ* plane. Click on **MIRROR** from the **CREATE** drop-down arrow, and select the head support body and the *YZ* plane. Click **OK** to complete the command.

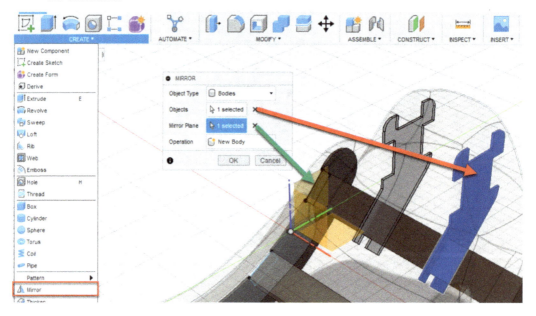

Figure 13.52 – The mirror head supports

5. To add in the supports, we need to add some lines that we cut from the previous sketch. Turn on the sketch that we just created (Fusion 360 turns off sketches by default), right-click on it, and select **Edit Sketch**.

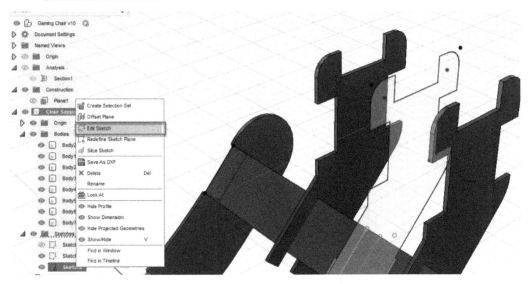

Figure 13.53 – The Edit Sketch tool location

6. Add three solid lines where the cut openings are located. Click on **FINISH SKETCH**. Note that Fusion 360 remembers the last profile that was selected and does not add these lines to close up the cutouts.

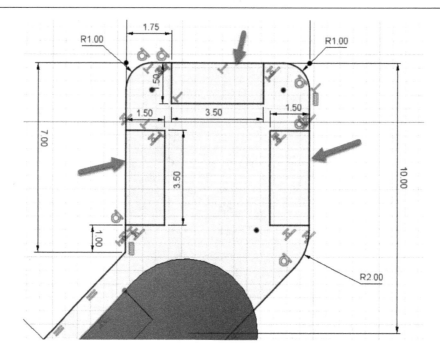

Figure 13.54 – Adding more lines to the sketch

7. Click on **Extrude** and select the three profiles that have now been created since we added those three lines. Set the direction to **Symmetric**, the extent type to **Distance**, the distance to 4.25 in, and the operation to **New Body**. Click **OK** to complete the command.

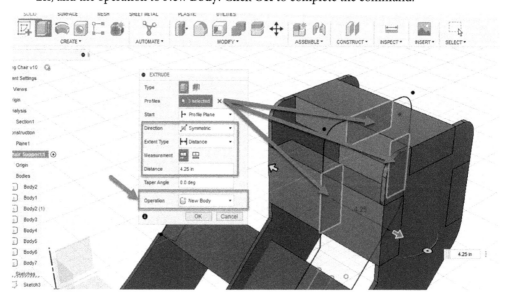

Figure 13.55 – Extruding from the center

8. To create the arms, we will create a plane at an angle and then offset that plane to create our sketch. Go to the **CONSTRUCT** drop-down arrow and select **Plane at Angle**. Pick the line edge that is second from the bottom of the 2x4 box. Set the angle to 15 and click **OK**.

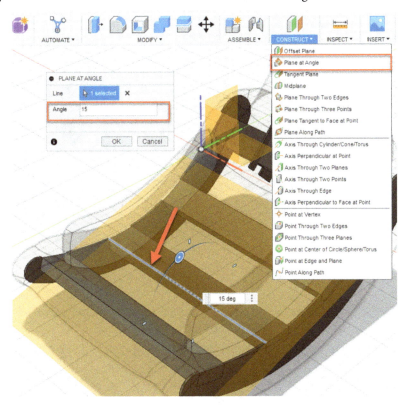

Figure 13.56 – Setting the plane at an angle

9. Go to the **CONSTRUCT** panel again, and this time, select **Offset plane**. Pick the previously created plane at an angle and then set the distance to 4.75.

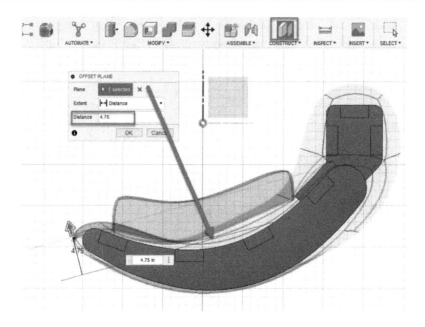

Figure 13.57 – Setting the offset plane

10. Click on the **Create Sketch** tool and select the offset plane that we just created. Draw a sketch like the one shown in *Figure 13.58*. Click on **FINISH SKETCH**.

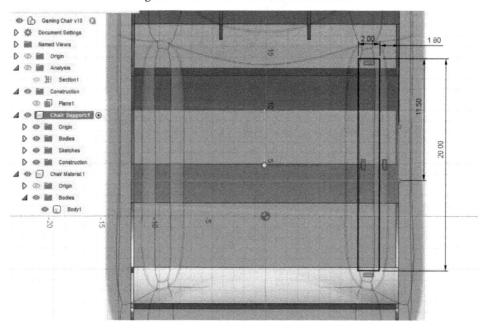

Figure 13.58 – Creating the arm sketches

11. Click on **Extrude** and select the profile that was just created. Extrude a distance of -7.5 `in` and create a new body. It is OK if it goes through the other rails for now; we will fix that in the next step. Click **OK** to finish the command.

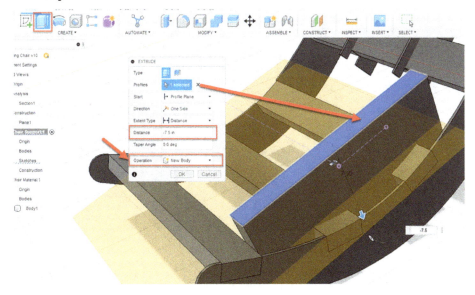

Figure 13.59 – Extruding to create an armrest

12. Click on **Create Sketch** and select the outside face of the body we just created.

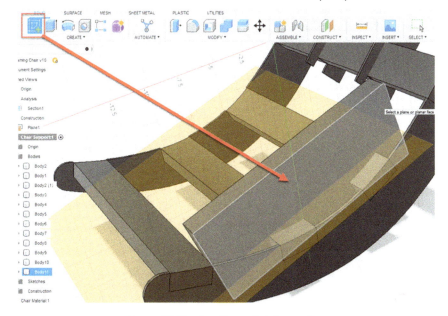

Figure 13.60 – Creating a sketch on a face

13. Turn off the side rocker legs to see the rail supports underneath. Click on the **CREATE** drop-down arrow, select **Project/Include**, and then select **Project**. Pick the two faces of the supports and click **OK**.

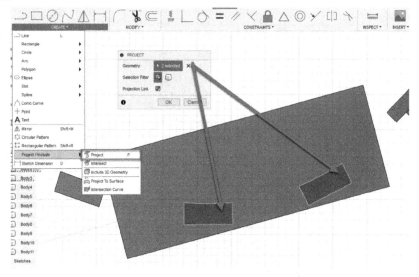

Figure 13.61 – Projecting the existing geometry to the current sketch

14. Draw two lines down from the corner points and set the constraints to colinear. A quick way to do this is to draw the lines from the bottom-corner points at varying angles, then add the colinear constraint, and then use the trim tool to remove the remaining lines. Do this for both rectangles. Click **FINISH SKETCH** when complete.

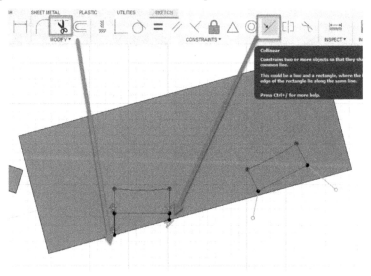

Figure 13.62 – Adding line geometry to the existing profiles

15. Orbit to see the bottom edges and turn off the support bodies to better see the sketch profiles. Select the **EXTRUDE** command, select the four sketch profiles, set the extent type to To **Object**, and select the opposite face of the sketch plane. Set the operation to **Cut** and click **OK**.

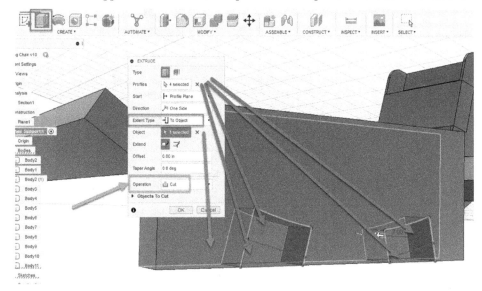

Figure 13.63 – Creating notches using the extruded cut tool

16. The final step is to mirror the body to the opposite side. Click on the **CREATE** drop-down arrow and select **Mirror**. Choose the arm body and then select the *YZ* plane. Click **OK** to complete the command.

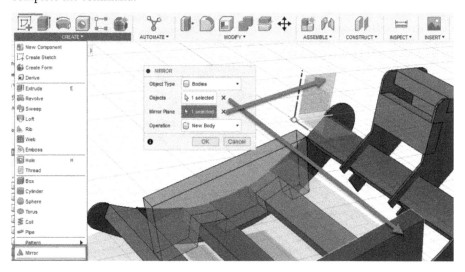

Figure 13.64 – Using the Mirror tool to create the opposite arm support

17. Turn on all bodies to see the result!

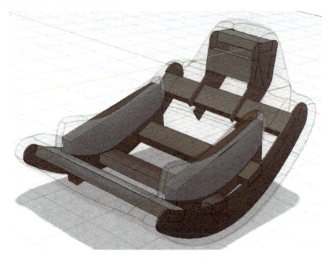

Figure 13.65 – The final interior framework

Improving the design

There is always room for improvement with any design, and this model is no exception. If you turn on the outer chair material, you'll see that the inner frame sticks through it a little bit at the top and the bottom. We have two options here; we can either adjust the frame or adjust the form. It would be easier in real life to adjust the form, as we could simply stuff that area with more padding, so we will adjust the form model to fix this issue:

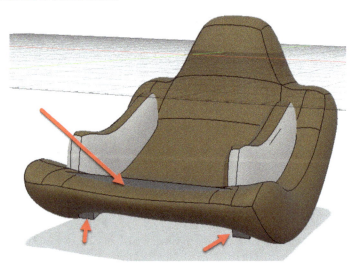

Figure 13.66 – The issues that need to be adjusted

1. Right-click on the **Form** icon in your Timeline, which should be the first thing that we created, and select **Edit** to go into the Form environment.

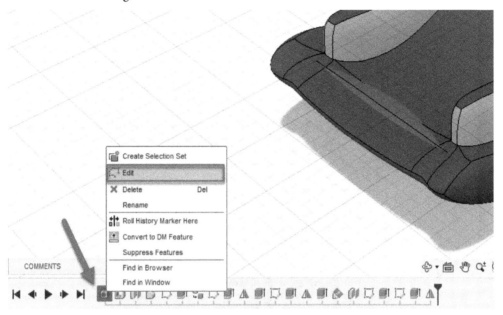

Figure 13.67 – Right-click to edit the original form

2. Turn on symmetry by going to the **SYMMETRY** drop-down arrow and selecting **Mirror - Internal**. Select the two opposite faces of the center line.

Figure 13.68 – Turning on symmetry

3. We will need to edit this in two steps since we don't want the geometry to warp. Select **Edit Form** in the **MODIFY** panel, and select the face shown in *Figure 13.69*. Drag the *Z* directional arrow in the positive direction about . 7 units.

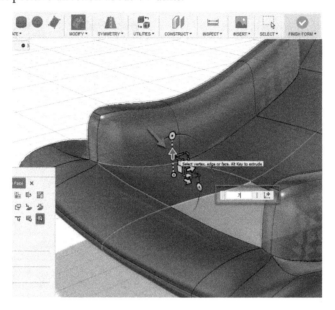

Figure 13.69 – Using Edit Form to move face geometry

4. Without closing the **Edit Form** tool, select the front face and move that in the *Z* direction about . 5 units. You can see that this now solves the frame issue. Click **OK** to complete the command.

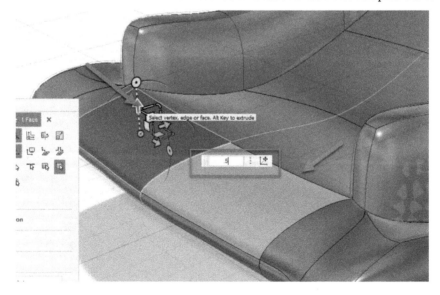

Figure 13.70 – Moving faces in the z direction

5. For the underside, we can select an intersection point and drag that over the frame. Select the point that is at the intersection of the frame and chair material. Drag the point in the negative direction about −2.4 units.

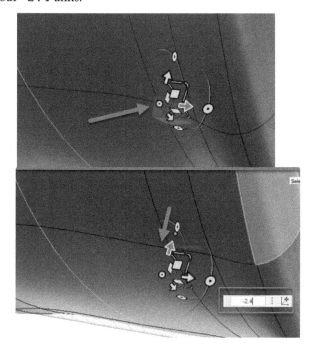

Figure 13.71 – Moving a point in the z direction

6. Select the intersection point next to the last selected point and move that down about −5. Click **OK** to close the **Edit Form** tool.

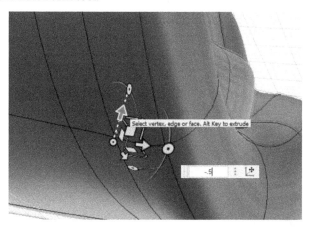

Figure 13.72 – Moving a second point in the negative Z direction

7. You may have noticed that the chair is under the grid. This was because when the donut tool was used to create the chair, the half that was not erased was underneath the grid. If I were to try to move the chair above the grid now, the interior would stay. We could keep it in this location, but it may help other users who may want to use your model to place it above the gridline, as it makes it easier to work with. Select the **Edit Form** tool and select the entire chair. Drag the chair in the positive Z direction about 25 units. Click **OK**, and then click on **FINISH FORM** to be taken back to the DESIGN environment.

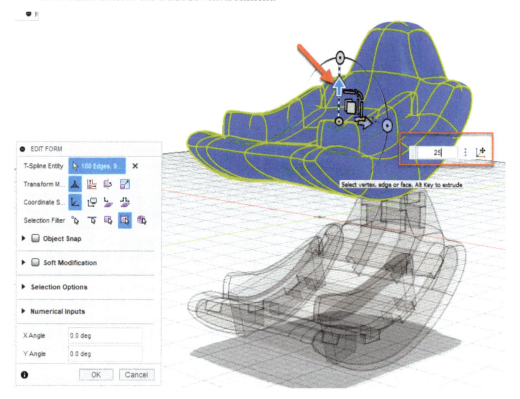

Figure 13.73 – Moving the chair material above the grid

8. As you can see, the interior framework stays where it is located. In order to move this up as well, we need to move the first sketch that started the model. Right-click on the first sketch in your Timeline and select **Edit Sketch**.

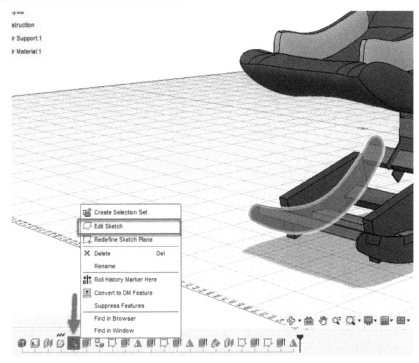

Figure 13.74 – Editing the first sketch in the Timeline

9. Note that the first sketch we created, which was the rocker leg support, is below the gridline and not connected to anything. We need to move this above the grid by the same distance as we did the chair material (25 units). Click on the **MODIFY** drop-down arrow and select **Move/Copy**.

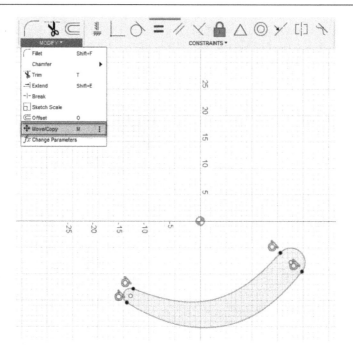

Figure 13.75 – The Move/Copy tool location

10. Drag a window around the entire shape, select the *Y* distance, and set it to 25 inches. Click **OK** to complete the move.

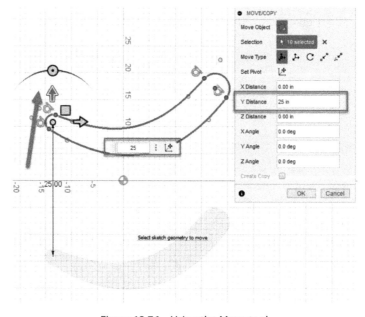

Figure 13.76 – Using the Move tool

11. Add dimensions using the *D* shortcut key and accept the default numbers. We don't want anything to move now. Click **FINISH SKETCH** when complete.

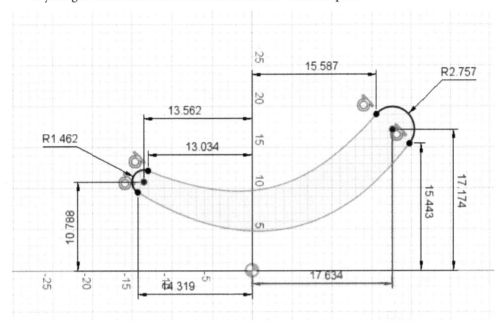

Figure 13.77 – Adding dimensions to the sketch

12. The chair is now above the gridline! This is because the rest of the sketches, planes, and so on further down the Timeline were based on this sketch. The Timeline is a very powerful feature within Fusion 360, and if you can master it, you can be a much better designer.

Figure 13.78 – The chair model is now above the gridline

13. There is one final adjustment that is needed. The arm supports wouldn't be a solid block of wood. Let's change this to make it a better design. To find this part, open the Chair Supports folder and then the Bodies folder drop-down arrow, and hover the mouse over the bodies until you see the arm model highlight in the view. Left-click on it, and note that it is highlighted in blue and shows up in the Timeline at the bottom, with three lines above it. The sketch before it is the sketch that created the body. Right-click on that sketch and click on **Edit Sketch**.

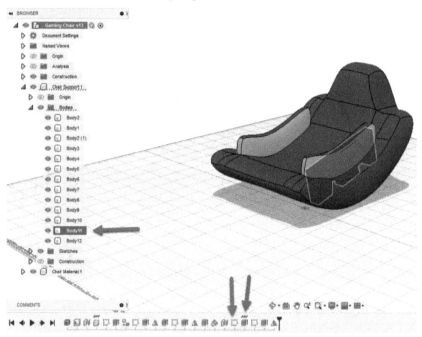

Figure 13.79 – Locating a body within the model and the Timeline

14. Adjust the dimensions according to *Figure 13.80*. Click **FINISH SKETCH** to finish.

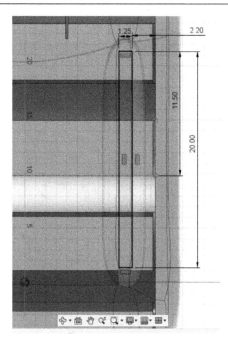

Figure 13.80 – The adjusted armrest dimensions

15. To make further adjustments, we need to first turn off the chair material to see the framework, and we will also need to drag the Timeline slider back to before the **Mirror** feature. Now, right-click on the sketch that created the cutaways for the arm.

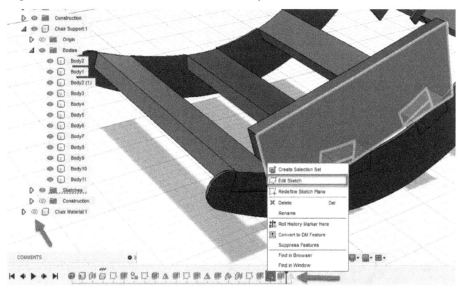

Figure 13.81 – Hiding the chair material and modifying the Timeline

16. We are going to remove some material to make this arm much lighter. Click on the **MODIFY** panel and select the **Offset** tool. Select the outer edge and set the distance to -0.75 in. It may also help to turn off the rocker legs to see the selection. Click on **OK** to finish with the offset tool.

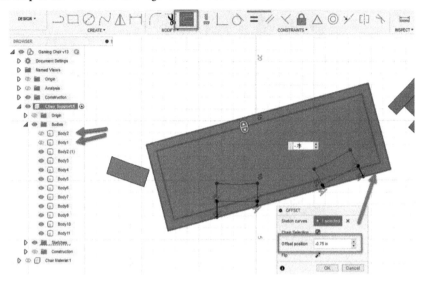

Figure 13.82 – Hiding the chair material and modifying the Timeline

17. Use the line tool to create an outer shape around the support legs since we need to add some material around them. Now, select the parallel constraint and select two lines to make them parallel (see *Figure 13.83*).

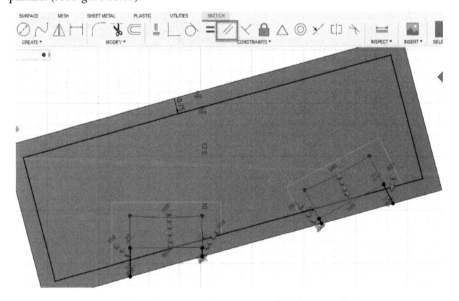

Figure 13.83 – Adding lines around supports and adding parallel constraints

18. Add 0.75-inch dimensions all around to set the wood thickness. Click on **FINISH SKETCH** to complete the sketch.

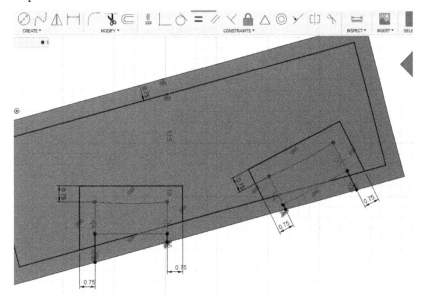

Figure 13.84 – Adding dimensions to a sketch

19. Right-click on the **Extrude** feature that created the cutaway and select **Edit Feature**.

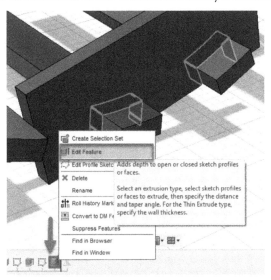

Figure 13.85 – Selecting the Edit Feature tool

20. The added sketch we created now shows up. Left-click on the middle area shown in *Figure 13.86*. The **Edit Feature** tool recognizes the previous selection we made to create the cuts at the bottom and adds the new selection.

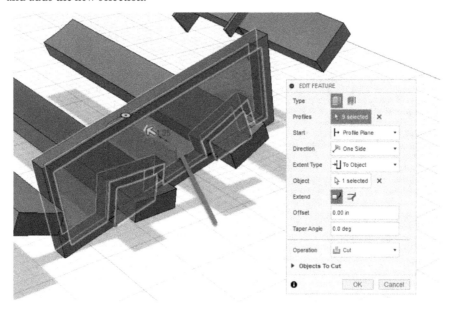

Figure 13.86 – Selecting the new profile and adding it to an existing feature cut

Changing the selected sketch plane

There is one final mistake that needs to be adjusted, which is that the sketch plane for the frame is on the wrong side and floats (see *Figure 13.87*). There is a quick fix for this, using a great tool that's useful when you accidentally select the wrong sketch plane:

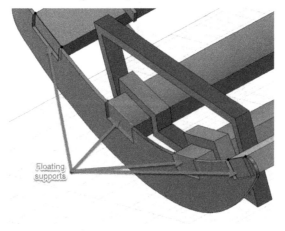

Figure 13.87 – The floating supports need to be fixed

1. Turn off the chair material (Body2 in my example) so that you can see the inner frame.

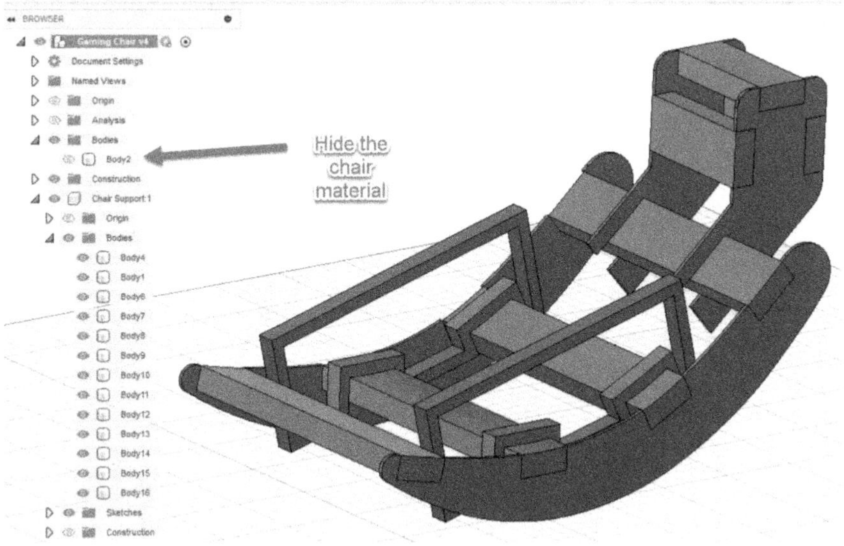

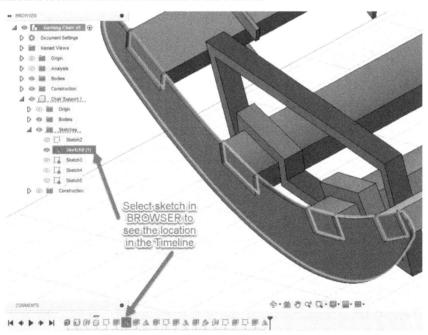

Figure 13.88 – Hide the exterior chair material

Turn on the sketch that was used to create the body supports, and left-click on that sketch in **BROWSER** to see where it is located within the Timeline at the bottom.

Figure 13.89 – Selecting the sketch to see it in the Timeline

1. Right-click the sketch that is highlighted in the Timeline, and select the **Redefine Sketch Plane** tool.

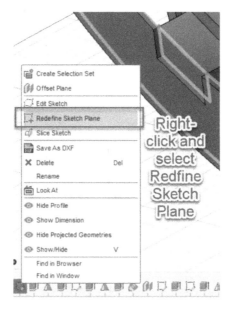

Figure 13.90 – Right-click and select Redefine Sketch Plane

2. Select the outside face of the support leg and click **OK**. The model will update, and the floating supports will now be fully on the support leg.

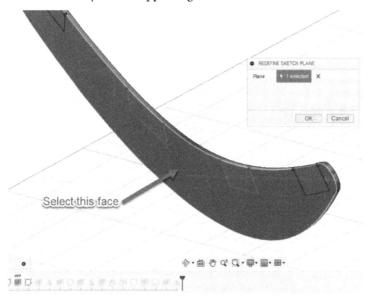

Figure 13.91 – Selecting the outer face

3. The floating legs have been fixed. Turn the chair material back on.

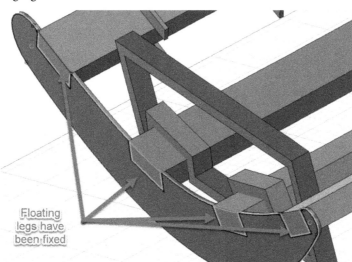

Figure 13.92 – The floating legs have been fixed

The chair is complete – at least according to me it is. Are there any changes that you would like to make? Go back in and adjust the sketches, the thicknesses, the outer form, the material, and so on until you are fully satisfied with your model.

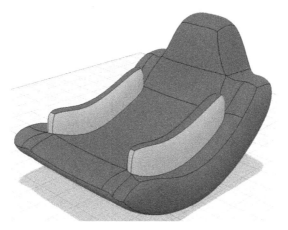

Figure 13.93 – The final chair model

I would love to see what you have created! Take a screenshot, add it to your favorite social media app, and tag my LinkedIn profile name.

Summary

In this chapter, we covered a lot of the Form environment and connected it to the DESIGN environment by creating an outer skin and an inner frame. We used the **Edit Form** tool to build the body, the **Section Analysis** tool to slice the model in half, and the **Shell** tool to hollow out the material, and we built a wooden frame interior with sketches. If you truly want to create this chair, you can bring all these parts into the Manufacture environment and add CNC paths to the parts. This would be best handled by a CNC expert, as knowing how to choose certain drill bits and how to best position these parts is beyond my expertise. In the next chapter, we will learn how to work with 2D- and 3D-scanned images.

Part 4: Working with 2D and 3D Scanned Images

In this final part, we will learn how to import images into Fusion 360 and scale those images up to full size so that the model can be created in Fusion 360. We will then finally create a bottle topper from a 3D scan and import that mesh data into Fusion 360 so that we can edit the model into a working part so that it can be 3D-printed.

This part has the following chapters:

- *Chapter 14, Using a Scanned Image to Create a 3D Model*
- *Chapter 15, Modeling a Bottle Topper*

14

Using a Scanned Image to Create a 3D Model

There are times when you may find something interesting online, such as a 3D model or a vehicle, and want to download it, but there are only pictures of the model online. This is where you will have to download the image, use it as a background image, and draw over the image to create a 3D model. We will learn how to use a photo of an online model (I have provided one for you) and then recreate it using Fusion 360. We will learn how to import it, scale it to the correct size, draw over the image, and recreate the 3D model. If you follow along with the tutorial video, be aware that some of the dimensions may not follow the book exactly, as the images were created at the time of writing.

In this chapter, we will cover the following main topics:

- Inserting and calibrating a scanned image into Fusion 360
- Adding sketch lines and dimensions
- Creating a 3D model

Technical requirements

You can practice with the files provided, or feel free to create your own for a more custom experience. The sample design for this chapter can be found at `https://github.com/PacktPublishing/ Improving-CAD-Designs-with-Autodesk-Fusion-360/tree/main/Ch14`.

Inserting a scanned image into Fusion 360

The image that we will work with is an ancient Asian sword that I found in the finest of the cheap, local dollar stores. We will learn how to import this image into our project folder and bring it into the design workspace:

Figure 14.1 – An Asian sword example

1. Let's start by opening Fusion 360 and starting a blank Fusion 360 design file:

    a. Click on the **Save** icon, create a new folder within the PACKT publishing project folder, and name it Ch14 Image Insert. Then, save it with the filename Image Insert.

2. Download the Sword.png image from the Ch14 GitHub folder link, as mentioned in the *Technical requirements* section.

3. Click on the **Data Panel** icon in the top-left corner of Fusion 360 to open your work folder location:

    a. Navigate to the Ch14 Image Insert folder and then click on the **Upload** button.

    b. Click on **Select Files**, locate the downloaded Sword.png image, and click on the **Upload** button.

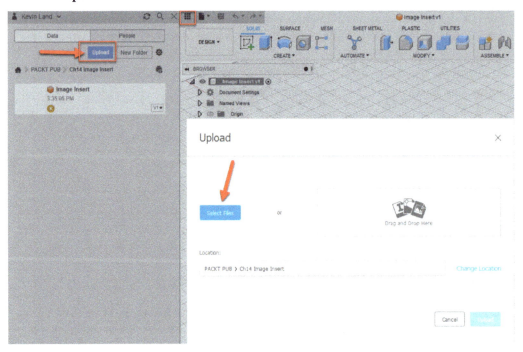

Figure 14.2 – The Data panel Upload icon location

> **Important note**
>
> You can upload all different types of files to your Fusion 360 project folder, such as PDFs, Word documents, and images. The project folder works the same as a typical Windows Explorer folder, so you can share your files with other designers on your team.

4. Now that the image has been uploaded into your work folder, let's place it on a plane:

    a. Go to the **INSERT** panel drop-down arrow and click on **Canvas**.

    b. The **INSERT** floating window should appear, showing the last folder location that we worked in, which should be Ch14. Click on the Sword.png image and then on the **Insert** button.

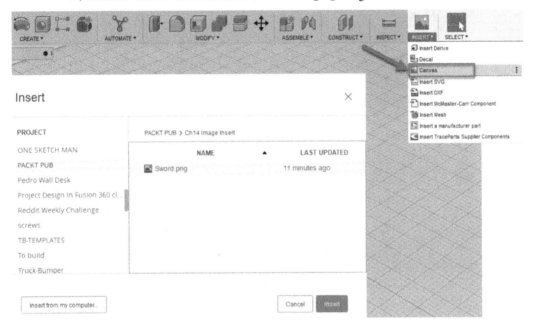

Figure 14.3 – The Canvas command location

5. Click on the XY plane, and the **CANVAS** floating window should appear with a variety of options:

    a. You may need to zoom in to see your image, as it may upload very small; don't change the scale right now, as we will use another tool to change that.

    b. Most of the options in the floating window you can leave as is, but the one that you definitely want to keep on is the **Display Through** option, as this allows you to see through your image, which will be helpful when we need to draw over the image.

c. We can always go back and adjust these options as well if we need to make any further adjustments later on, such as flipping the image to a different side. For now, click **OK** to place the image.

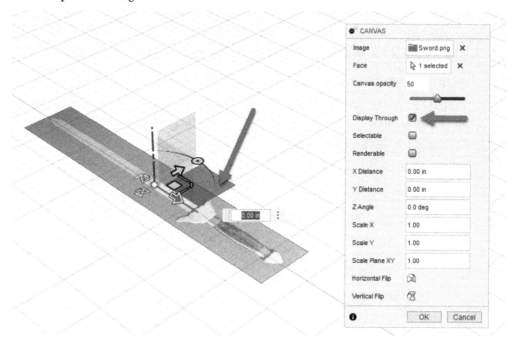

Figure 14.4 – The CANVAS window options

The image has now been placed onto the gridplane. We need to scale up the image, but the scale options within the **CANVAS** tool only scale from the middle outward. We want to calibrate from point to point, which we will learn to do next.

Calibrate the image to size

The **Calibrate** option allows us to select two points on an image and scale by those two known dimensions. This works great when you know the general size of an object's length but not many other dimensions are given. Let's look at the **Calibrate** tool now:

1. Go to the **BROWSER** window on the left side of your screen and open the Canvases folder, which was added when the image was placed:

    a. Right-click on the **Sword** image and then click on **Calibrate**. Nothing will show up on the screen yet.

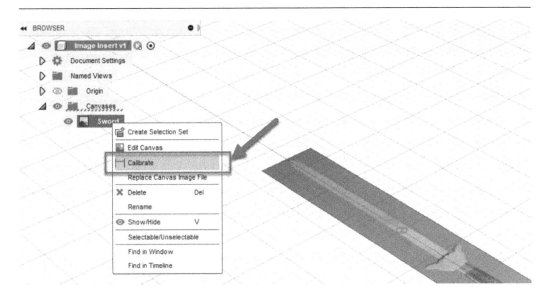

Figure 14.5 – The Calibrate tool location

2. Click on the **TOP** view of the ViewCube to see the sword image from above:

    a. Hover the mouse over the handle location and click on the top part of the handle.

    b. A green dot appears, asking you to place another location. Click on the bottom of the handle to place a second dot, and then a dimension will appear.

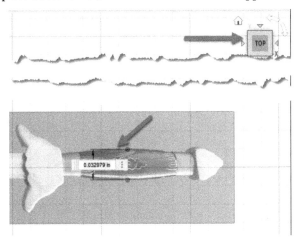

Figure 14.6 – The Calibrate tool two-point dimension location

3. Since there are no dimensions known about this sword, we can guess that the handle thickness is about 2 inches thick. Change the dimension to 2 inches and hit *Enter*.

Zoom out to see that the sword has now been calibrated to be much closer to real-life size. Now that the image is full-size, we can add sketch lines to start generating geometry.

Adding sketch lines and dimensions

We can now add the sketch geometry to generate the 3D model. Let's first adjust the location of the image so that the tip of the sword is placed close to the origin dot. The reason why we will want to move the image is because it will give us a better center location if we want to mirror the sword on the opposite side:

1. Open the `Canvases` folder in the **BROWSER** window on the left-hand side, and then right-click on the **Sword** image. Then, select **Edit Canvas**.

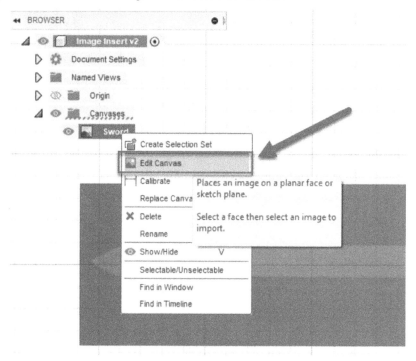

Figure 14.7 – The Edit Canvas tool location

2. With the **EDIT CANVAS** window open, we can now move the image:

    a. Drag the **X Distance** arrow by about `20.4` inches.

    b. Then, drag the **Y Distance** arrow in the negative direction by `-.13` inches to place the tip of the arrow close to the origin dot; it may not be exact, but it's close to what we need. Click **OK** to complete the command.

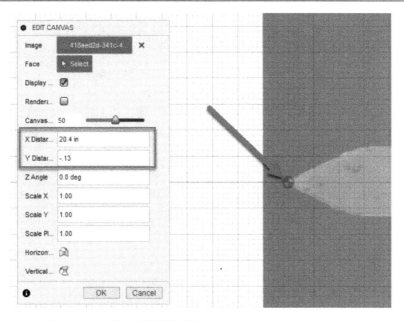

Figure 14.8 – The EDIT CANVAS tool options to move the image

Let's separate this sword into two parts, such as the sword and the handle. This way, it can be built into two separate parts if you want to 3D-print it. To do this the right way, we need to create two components:

1. Go to the **ASSEMBLE** panel and click on **NEW COMPONENT**. Change the name to Sword and click **OK**.

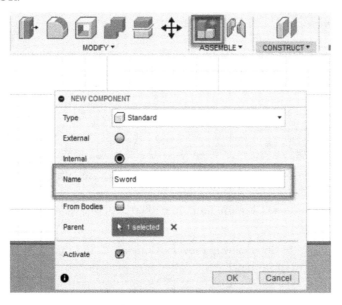

Figure 14.9 – The NEW COMPONENT tool location and name

2. Change the active component to the top-level assembly. This way, you won't create a multi-part component within the sword and the handle will be a separate part.

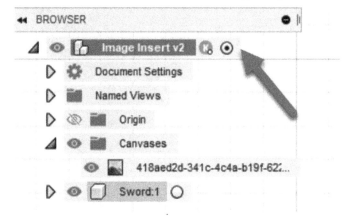

Figure 14.10 – The Active Component button location

3. Since our mouse is over this way, right-click on the top-level assembly name and select **New Component**. Name this new component Handle and click **OK**. Note that as you become more proficient with Fusion 360, the fewer mouse movements and clicks you will make, and the faster you will become.

Figure 14.11 – The New Component right-click tool location

4. Select the **Sword Activate Component** button, then click on **Create Sketch**, and then select the *XY* plane.

5. Draw a sketch using the line tool with the same dimensions and constraints as in *Figure 14.12* (I've cut the image to make viewing the dimensions easier for you). Click on **Finish Sketch** when you are done.

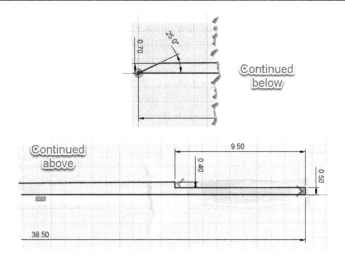

Figure 14.12 – The sword sketch dimensions and constraints

6. Instead of immediately extruding this part right away, let's create the sketch for the handle. This way, we can capture the sketch data from the sword much easier without the 3D model. Activate the **Handle** component and create a sketch on the *XY* plane.

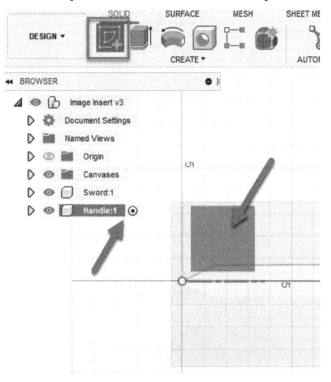

Figure 14.13 – Activating the Handle component

7. Create a shape like the one shown in *Figure 14.14* for the sword guard. Note that I'm only creating half of the sword and half of the handle. This is typical when using design programs; if you see an object that has two of the same sides, use the mirror tool to make your design much easier and faster to create. These dimensions are also for your reference. If you have a better shape that you want to go with, then change your design as you wish.

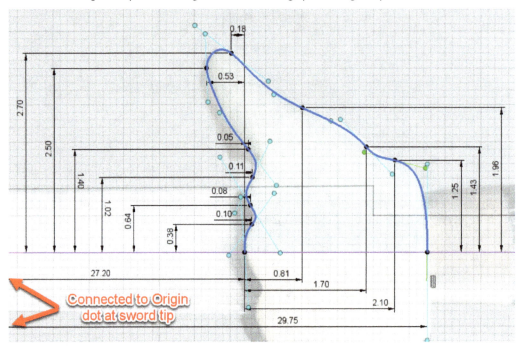

Figure 14.14 – Creating the shape of the sword guard

Important note

If it's hard to see the image below while you are tracing, go into the `Canvases` folder, right-click on the image, and select **Edit Canvas**. Then, increase the canvas opacity slider until the image is much easier to see.

---

8. In the same sketch, pan over to the hilt of the sword. We can either create another sketch or use the same sketch; Fusion 360 is not picky:

    a. Create a drawing similar to the one shown in *Figure 14.15*. Note that this part of the image has been warped a bit by the camera lens. We have a bit more creative freedom here. Try your best to keep the dimensions close to rounded-off numbers; this way, it makes it easier for you to move things later. Click on **FINISH SKETCH** to complete the hilt and guard.

For the handle, we will use the loft tool to generate the shape between the guard and the hilt.

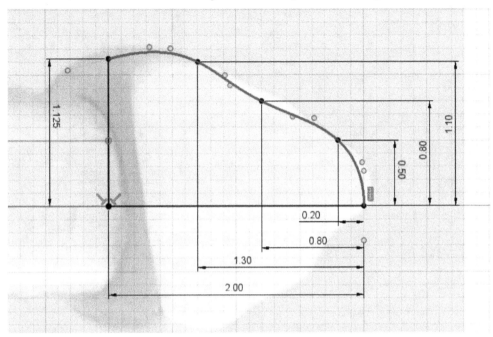

Figure 14.15 – Creating the shape of the sword hilt

Now that we have a general 2D sketch shape, it's time to generate our 3D models from our sketches.

Creating a 3D model

Now, we can start to create a 3D model of the sword and handle. We can use Google to determine the thickness of the sword and the general size of the rest of the handle.

The sword model

We will create the 3D model of the sword first to give us a guide for the centerline of the handle, which we will create next. Follow these steps to create the sword:

1. After doing a quick Google search, I found that a typical sword thickness is about .28 inches thick. So, let's extrude the sword to that thickness:

    a. First, set the sword as the Active Component, then click on the **EXTRUDE** tool, and select the sketch profile for the sword.

    b. Set **Direction** to **Symmetric** so that the sword is extruded on both sides of the sketch profile, which will make it easier to mirror along that profile.

c. Set **Measurement** to the whole length and **Distance** to .28 inches, and then click **OK**.

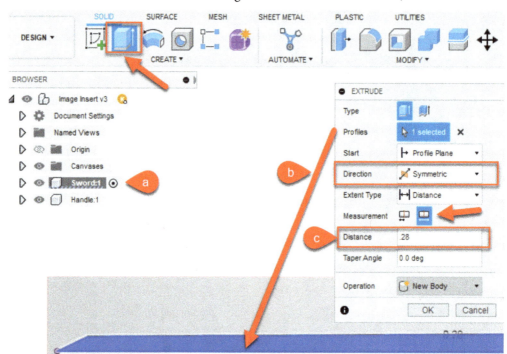

Figure 14.16 – Extruding the Sword profile

2. To create the sharp top edge, we will use the **Draft** tool:

a. Click on the ViewCube and orbit it to view the left side.

b. Click on the ViewCube rotation arrow, rotate the cube 90 degrees clockwise, and then orbit the view slightly to see the bottom of the 3D model.

c. Go to the **MODIFY** drop-down arrow and select the **Draft** tool.

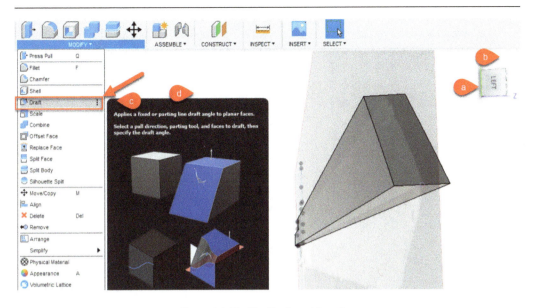

Figure 14.17 – The Draft tool location

3. Select the pull direction to be the bottom of the sword, and then select each outside face for the faces that will receive the draft angle. Now, set the draft sides to be **Symmetric** and the angle to be 11 degrees. Click **OK** to complete the command.

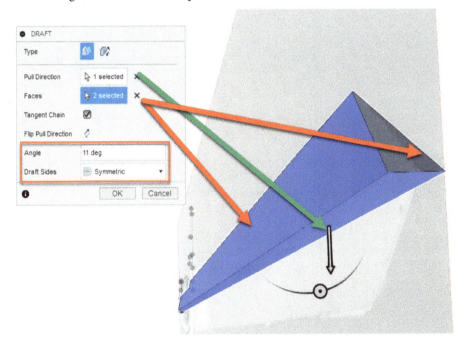

Figure 14.18 – The Draft tool options

4. We have a nice-looking side edge of the sword, but the tip has too much width. We need to shape the sword into a nice, sharp point. Click on the **Create Sketch** tool and select the top face of the tip.

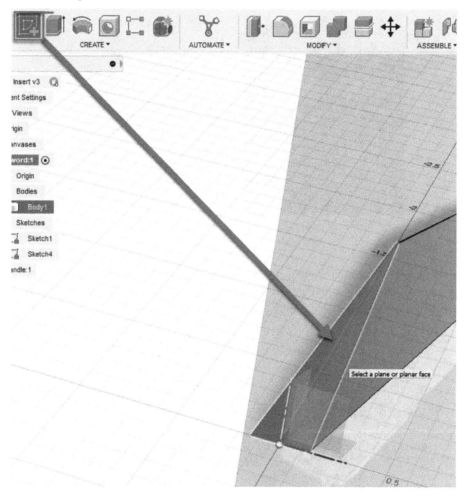

Figure 14.19 – Creating a sketch on the sword tip face

5. We will create a profile to cut some extra material away from the tip:

    a. Draw two lines from each edge to the outside of the sword (see the green arrows in *Figure 14.20*).

    b. Use the **Project** tool to select the outer edges (the red arrows).

    c. Finally, draw a line from the tip of the sword back to the rear corners (the orange arrow).

    d. Click on **FINISH SKETCH**.

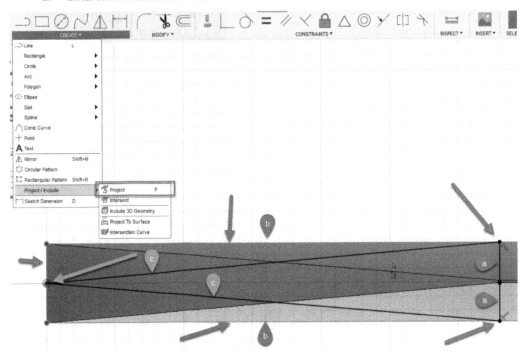

Figure 14.20 – The sketch lines to sharpen the sword tip

6. Click on the **EXTRUDE** tool, select the outer extrude profiles, and set **Extent Type** to **All** and **Operation** to **Cut**.

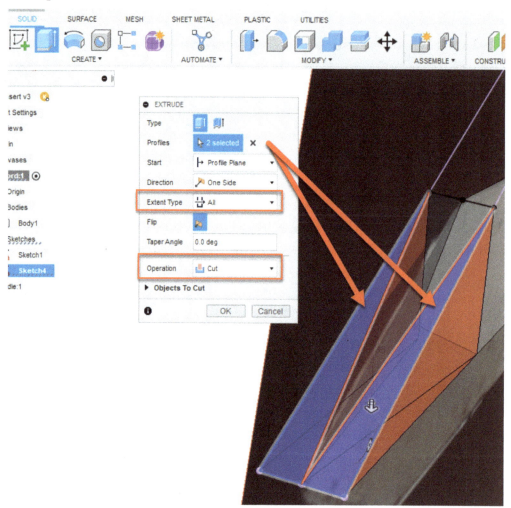

Figure 14.21 – The EXTRUDE tool options

7. We can now mirror the sword to create the final shape. Click on the **CREATE** drop-down arrow and select **Mirror**. Select the sword body, and then select the *XY* plane to mirror the body. Set **Operation** to **Join** and click **OK**.

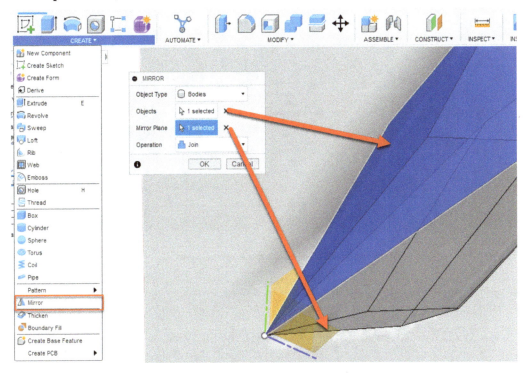

Figure 14.22 – The Mirror tool location and options

The sword section is all set now and looks great! We can work on the handle guard and the hilt next.

The guard and hilt

We can now extrude the guard and the hilt as two separate bodies and then give them some added shape:

1. Let's hide the sword first to get it out of the way:

   a. Turn off the **Sword** eyeball and activate the **Handle** component.

   b. Click on the **EXTRUDE** tool and select the **Guard** profile.

   c. Set **Direction** to **Symmetric**

   d. Set **Measurement** to the whole length.

   e. Set **Distance** to 1 inch and click **OK**.

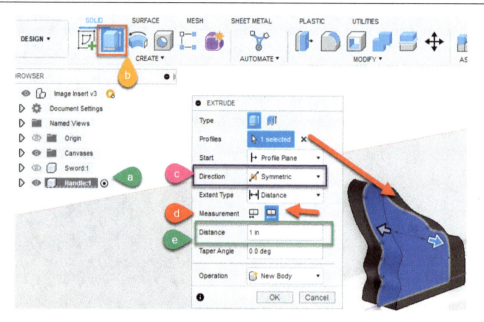

Figure 14.23 – Extruding the sword guard

2. The guard has been extruded, but it looks a bit too square. Let's give the handle a curve and then smooth out the edges with a fillet:

    a. Click on the **Create Sketch** tool and then click on the *YZ* plane.

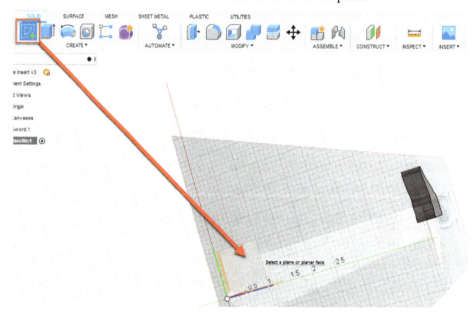

Figure 14.24 – Create Sketch on the YZ plane

3. Select the **PROJECT** tool, set **Selection Filter** to **Body**, and then select the sword guard. We need to project this body, as we want to grab the midpoint of the edge, and since this edge is a curve, it needs to be projected in order to be selected.

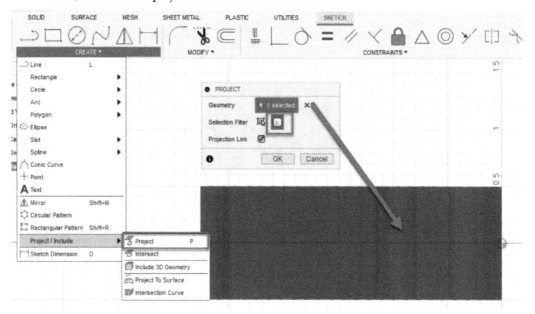

Figure 14.25 – Projecting the guard body

4. Let's create the shape we will use to add a curve to the guard:

   a. Click on the line tool and draw a horizontal construction line from the origin dot to the left midpoint edge.

   b. Click the **Spline** tool and select the midpoint edge and the top corner point.

   c. Use the **Fit Point** spline handles to create a spline (as shown in *Figure 14.26*) and create a smooth curve.

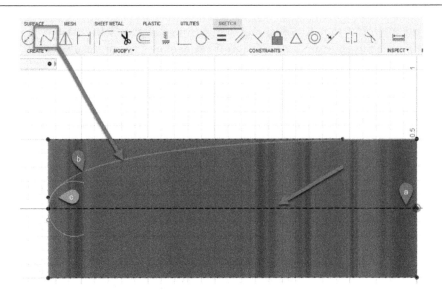

Figure 14.26 – Using the construction line and spline tool

5. Let's mirror that spline at the bottom so that the smooth cut happens on both sides:

    a. Click on the **CREATE** panel drop-down arrow and select the **Mirror** tool.

    b. Select the spline as the objects.

    c. Then, click on the construction line as the mirror line. Click **OK** and then click **FINISH SKETCH**.

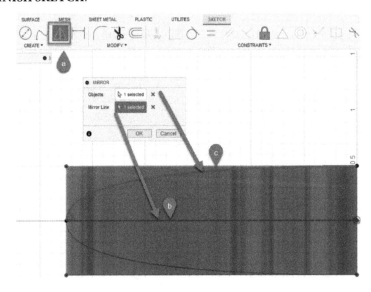

Figure 14.27 – The Mirror tool on the curved spline

6. Let's cut out that spline by using the **EXTRUDE** tool:

    a.   Click on the **EXTRUDE** tool and select the area above the spline, which is what will cut the area away from the guard.

    b.   Select **Extent Type** to **All**.

    c.   Set **Operation** to **Cut** and click **OK**.

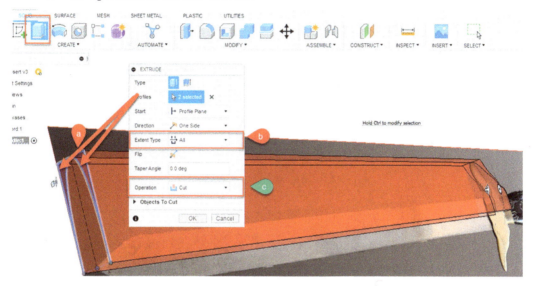

Figure 14.28 – An extruded cut to add curve to guard body

7. We have a nice, curved guard body half now. We will add some more fillets after we mirror to smooth it out more, but first, we need to mirror to the other side:

    a.   Select the **CREATE** panel and click on the **Mirror** tool.

    b.   Select the body as the objects.

    c.   Select the bottom guard face as the mirror plane.

    d.   Set **Operation** to **Join** and click **OK**.

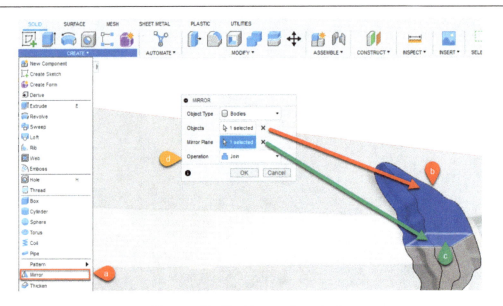

Figure 14.29 – The Mirror tool on a body

8.   Let's repeat the process for the sword hilt:

   a.   Turn on **Sketch1** (your number may vary) and select the **EXTRUDE** tool.

   b.   Pick the hilt profile and set **Direction** to **Symmetric**.

   c.   Set **Measurement** to the whole length.

   d.   Set **Distance** to 1 inch and click **OK**.

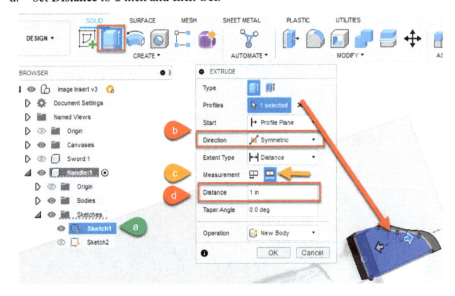

Figure 14.30 – Extruding the sword hilt

9. We will add a nice smooth cut to remove the hard edges, as we did for the sword guard. Select the **Create Sketch** tool and select the flat edge of the sword hilt. Previously, we couldn't select this edge, since the guard had a lot of curves, but since this is a flat face, we can create a sketch on it.

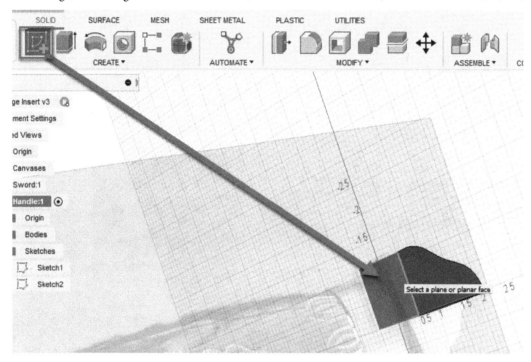

Figure 14.31 – Creating a sketch on a face

10. Select the **Project** tool, the body selection filter, and the hilt body, and then click **OK**.

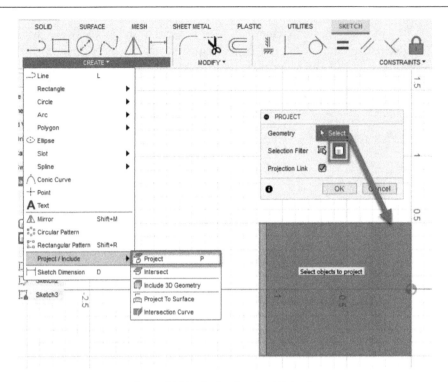

Figure 14.32 – Projecting the hilt body

11. Add a horizontal construction line, and then add a spline curve that attaches to the top but comes in slightly from the left edge (see the orange arrow in *Figure 14.33*). The reason that we want to do this is because it will allow for smooth geometry for the cut operation.

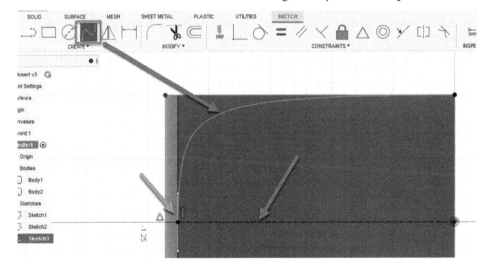

Figure 14.33 – Creating a construction line and spline curve

12. Mirror the spline to the bottom side by going to the **CREATE** panel, selecting **MIRROR**, and choosing the spline as the object and the construction line as the mirror line. Click **OK** to complete the command and then click **FINISH SKETCH**.

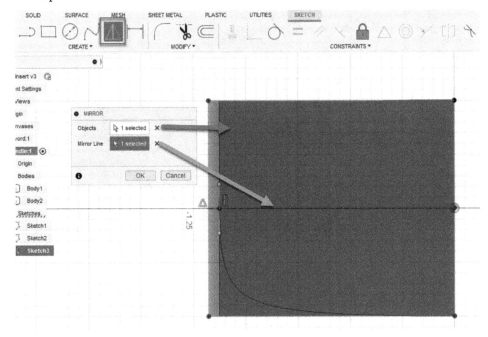

Figure 14.34 – Mirror the spline curve at the bottom

13. Add an extruded cut to remove the outside hard edges to add the smooth curve. Click on **EXTRUDE** and choose the three profiles, then set **Extent Type** to **All** and **Operation** to **Cut**, and click **OK**.

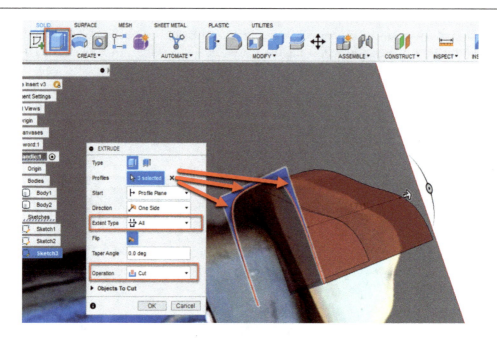

Figure 14.35 – Adding an extruded cut

14. Click on the **CREATE** drop-down arrow and select the **Mirror** tool. Select the body and then the bottom face as the mirror plane. Click **OK** to finish the command.

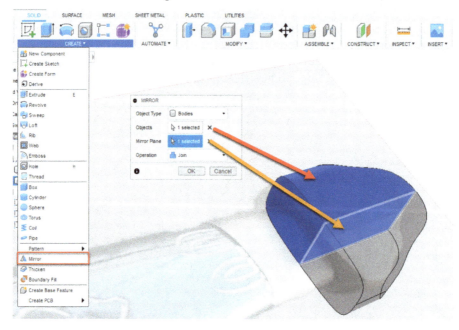

Figure 14.36 – Adding the Mirror command

15. To view everything we've done so far, turn off the **Canvas** eyeball icon and click the top-level assembly to view all the components.

Figure 14.37 – The sword model so far

The sword, guard, and hilt all look great so far. Next, we will use the **Loft** tool to create the handle, then add some fillets to smooth out the body, and add some nice finishing touches to complete the sword.

The handle

To create the handle, we will create three sketches on offset planes and then use the **Loft** tool to connect them all together:

1. Make the Active Component handle and then zoom into the handle area. Click on **Create Sketch** and select the flat face on the hilt.

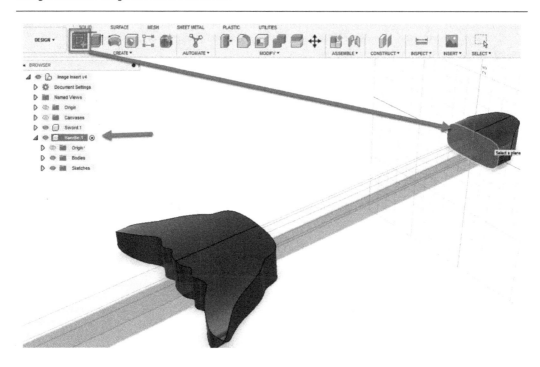

Figure 14.38 – Adding a sketch to the hilt face

2.   Turn off the Sword folder main eyeball icon and turn off the eyeball icon for **Body1**, which is the sword guard. This will allow us to see the sketch face that we will work on. Click on the **CREATE** drop-down arrow and select the **Ellipse** tool. Click on the center origin dot and create an ellipse that is 1.75 inches by .85 inches. Click **FINISH SKETCH** to complete the sketch.

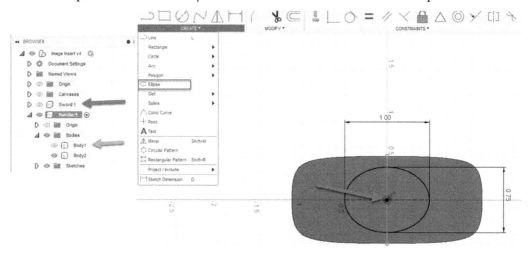

Figure 14.39 – Adding an ellipse to the hilt face

3. Turn on **Body1**, which should be the guard. Orbit to see the interior face facing the hilt. We need to create a sketch on that face, but if you try to select that face to draw on, you'll notice that we are unable to select it. This is because Fusion 360 can only create sketches on flat surfaces. We need to create a surface to draw in that location. First, we need to know the distance at which to create our offset:

a. Click on the **MEASURE** tool in the **INSPECT** panel and select the two top points, and a temporary distance of 8.75 inches will be displayed. You can click on any face, edge, or point to get a distance between two locations. Click on Close to complete the command.

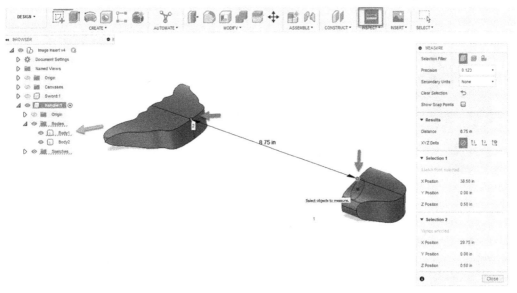

Figure 14.40 – Checking a distance with the INSPECT tool

4. Now that we know the distance to place the offset, we can use the offset plane tool to create the sketch on. Click on the **CONSTRUCT** drop-down arrow and select **Offset Plane**. Click on the flat hilt face where we created the previous sketch and set **Distance** to 8 . 8 inches, just a few more than 8.75 inches. The reason why we don't set it to 8.75 inches is because the surface we are creating is slightly curved, and we need just a few more inches to make sure the loft does not have any gaps around the edges.

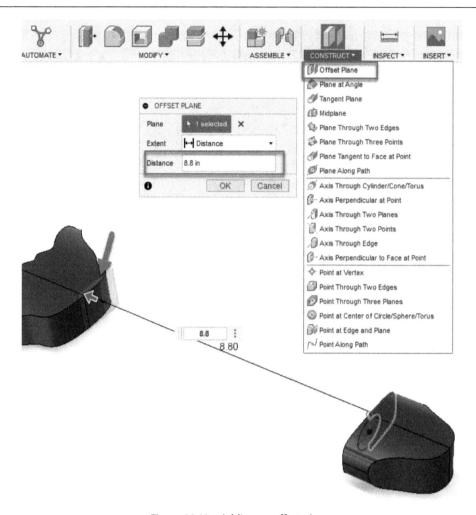

Figure 14.41 – Adding an offset plane

5.   Click on the **Create Sketch** tool and select the offset plane that we just created.

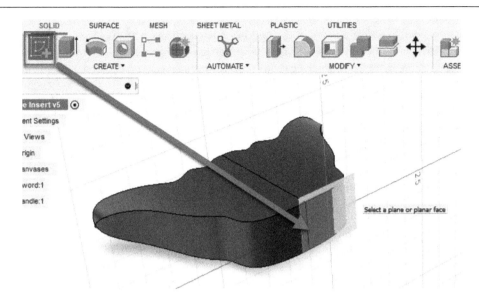

Figure 14.42 – Creating a sketch on the offset plane

6. We need to create another ellipse on this new construction plane. To make things easy for us, we will project the previous sketch into this new sketch. Turn off **Body2** within the `Bodies` folder to see the new sketch. Select the **CREATE** drop-down arrow and hover the mouse over the **Project / Include** arrow to select the **Project** tool. Click on the ellipse that was created in the previous sketch to project that sketch into this sketch.

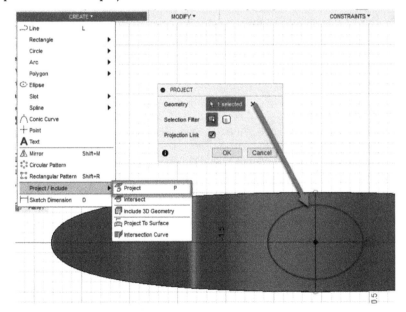

Figure 14.43 – Projecting the ellipse from a previous sketch

7. We now need to create a plane in the middle, since the handle grows larger in the middle area. Turn back to **Body2**, as we will need a face to select and turn on **Plane1** within the `Construction` folder. Select the **CONSTRUCT** drop-down arrow and then select **Midplane**. Select the hilt face and then select the previous construction plane. Fusion 360 calculates the exact middle for you. Click **OK** to complete the command.

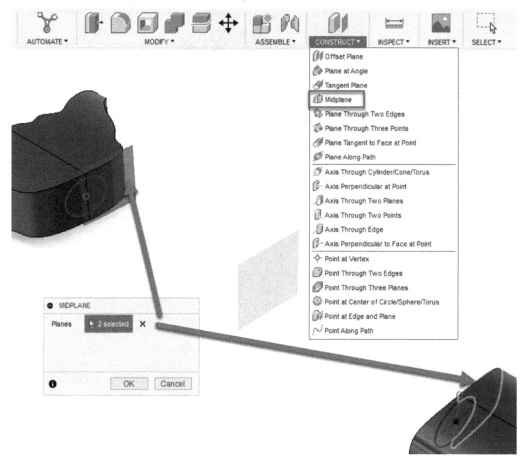

Figure 14.44 – Creating a midplane between two faces

8. Select the **Create Sketch** tool and select the new midplane.

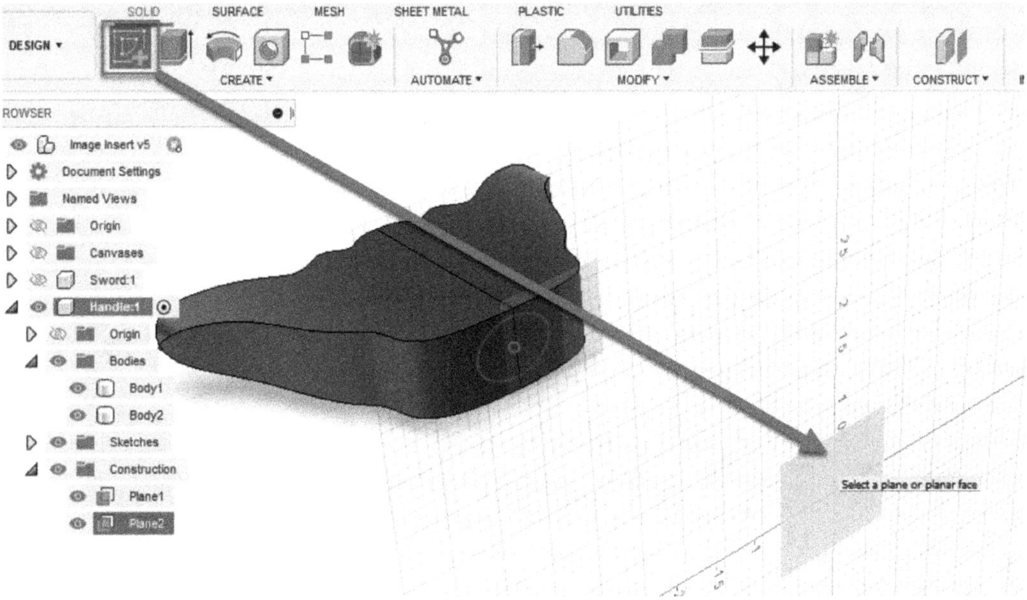

Figure 14.45 – Creating a sketch on the midplane

9. Note that something odd is happening with our design. It shows that the ellipse sketch that we previously placed on the origin dot is now no longer placed on that origin but is off to the left. It really didn't move at all. What happened here is that when we created the offset sketch from the hilt face, it created a new origin for that construction plane. Since the hilt face was split into two parts due to the ellipse in the middle (see *Figure 14.44*), Fusion 360 placed the origin of that construction plane in the middle of that face only. Be aware that Fusion 360 will create new origins for construction planes that are on different faces. Click on the **CREATE** drop-down arrow and select the **Ellipse** tool. Create an ellipse that is 2 inches by 1 inch in the center of the previous ellipse. Click on **FINISH SKETCH** to complete the sketch.

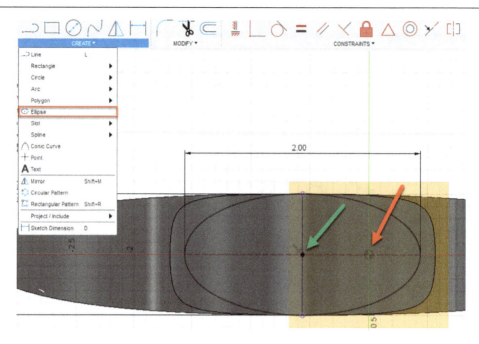

Figure 14.46 – Create the ellipse on the ellipse origin dot

10. Click on the **CREATE** drop-down arrow and select the **Loft** tool. Click on each profile to generate the loft body. If you have trouble selecting the body in the guard body, hold down the left mouse button for a few seconds over the profile that you are trying to select, and you can use the filter dropdown to choose the profile. For profile 2 and 3, you will also need to select both sides of the sketch, since it is split in two due to the purple projection line.

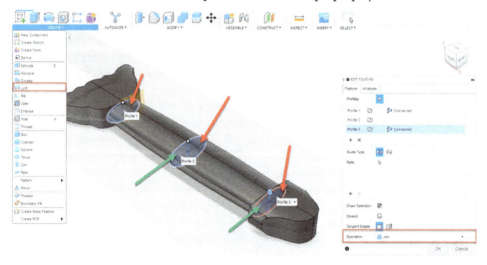

Figure 14.47 – Create a loft between three profiles

11. Turn on the sword body by clicking on the eyeball icon and scroll back to see the entire sword model.

Let's finish off the model with some finishing touches, such as fillets, to smooth sharp corners and some details in the sword handle.

The finishing touches

Let's add some details to the handle to make it stand out a bit:

1. Make sure the **Handle** component is active and click on the top view of the ViewCube. Turn on the **Canvases** eyeball icon as well to see the image underlay.

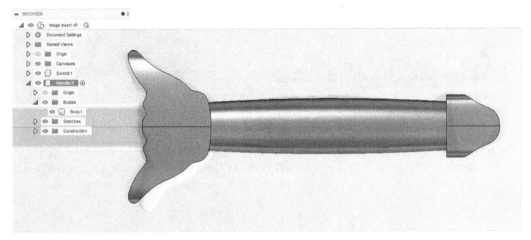

Figure 14.48 – The view from the top with the canvas image on

2. We will create the two gold end pieces that are just below the guard and the hilt. To quickly do this, we will use the **Split Face** tool. First, we need to add two offset planes, as we will use these planes as our cut planes. Click on the **CONSTRUCT** drop-down arrow and select **Offset Plane**. Select the flat face at the top of the hilt, set it to 1 inch, and then click **OK**.

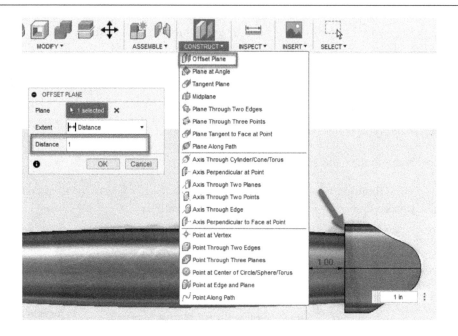

Figure 14.49 – Adding an offset plane from the hilt

3. Repeat the same process for the guard, but instead of selecting a face, select the construction plane. If it is turned off, go to the Construction folder, and turn on the eyeball icon.

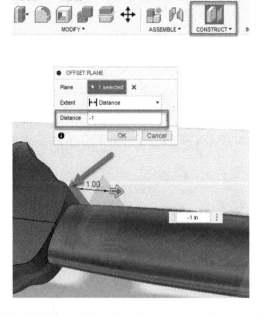

Figure 14.50 – Adding an offset plane from the guard construction plane

4. Now, let's create the splits. We will use the **Split Face** tool, as this is non-destructive to the body that is chosen, meaning that when we split the faces, it will add extra faces to an existing body. Don't get this confused with the **Split Body** tool, which is destructive and will result in the body slitting in two and another separate body being added to the `Bodies` folder in the browser. Go to the **MODIFY** drop-down arrow and select **Split Face**. The **SPLIT FACE** floating window will open. For **Faces to Split**, select the handle, and for **Splitting Tool**, select both construction planes, and then click **OK**. Turn off the `Construction` folder eyeball icon in the browser to see the splits in the handle.

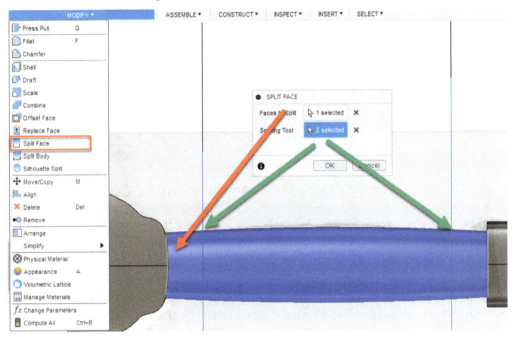

Figure 14.51 – The location of the Split Face tool and the selections

5. We will add the leaf detail (I think it is a leaf) to the hilt first and then use a path array to add more. First, let's create an offset plane from the *XY* plane, set **Distance** to `1.5` inches, and then click **OK**.

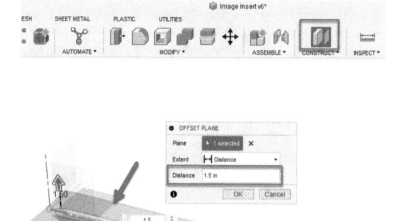

Figure 14.52 – Create an offset plane from the XY plane

6.   Create a sketch on that newly created offset plane. If you want to make the construction plane larger, click on a corner point and drag it to make it larger.

7.   Create a sketch like the one shown in *Figure 14.53*, using the **Project** tool to capture the body of the rear plate, and then use the **Fit Spline** tool to create the leaf shape. I created only half of the fit spline and then used the **Mirror** tool to create the opposite side, but feel free to use whatever tool you feel comfortable with. Click on **FINISH SKETCH** when done.

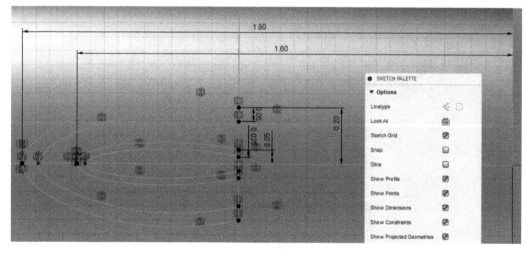

Figure 14.53 – Project the rear face and then draw a leaf sketch

8. Click on the **Extrude** tool, set **Distance** to −1.1 and **Operation** to **Cut**, and then click **OK**.

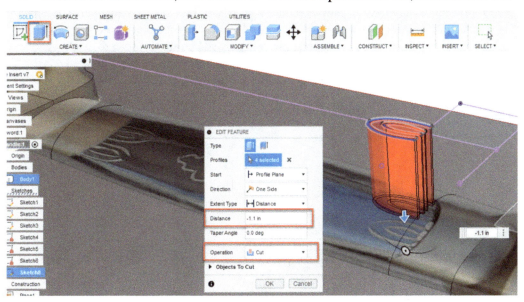

Figure 14.54 – Extrude the leaf sketch toward the handle by -1.1 inches

> **Important note**
> Fusion 360 has an Emboss tool that works great on cylinders but not so great on spline ellipse shapes, such as the one we have here, which is why we must use the **Extrude** tool.

9. Now, we need to create the path array that will pattern the extrusion that we just created around the ring:

    a. Click the **CREATE** drop-down arrow, select the **Pattern** arrow, and then select **Pattern on Path**.

    b. Click on the **Object Type** drop-down arrow, and set it to **Features** so that we can capture the extrusion feature that we made in the previous step.

    c. For **Objects**, select the **Extrude** feature in the **Timeline**.

    d. For **Path**, select the slice that we created earlier.

    e. Set **Quantity** to **2** and **Distance** to 0.5 inches, and then click **OK**.

Set **Distribution** to **Spacing** and **Orientation** to **Path Direction**; this way, the feature will space evenly and follow the path direction. Typically, if this were a circular curve, the path would go all the way around, as would the arrayed selected objects. The path does not go all the way around due to the odd curvature, so we will only add one array object to get the spacing location, and then mirror it to the opposite side. We will then mirror it to the guard object.

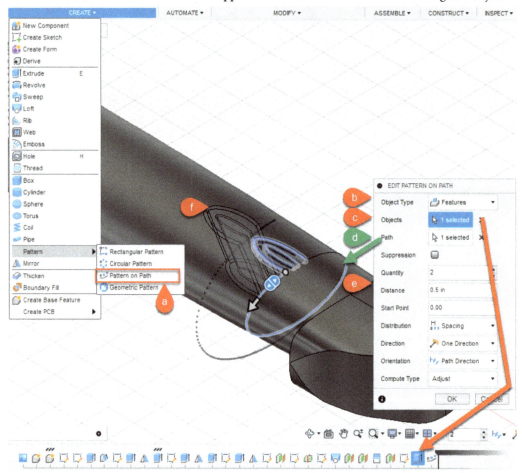

Figure 14.55 Pattern on Path tool

Important note

Fusion 360 is very finnicky when it comes to path arrays; sometimes it works and sometimes it does not. It likes nice clean 2D paths, not elliptical paths like the one we have, but we can trick it to work the way we want.

10. The cut has been placed at a slight angle. We will now mirror that array feature to the other side. Click on the **CREATE** drop-down arrow and select **Mirror**. Set **Object Type** to **Features**, and select the path feature that we created in the previous step. Select the *XZ* plane for the mirror plane.

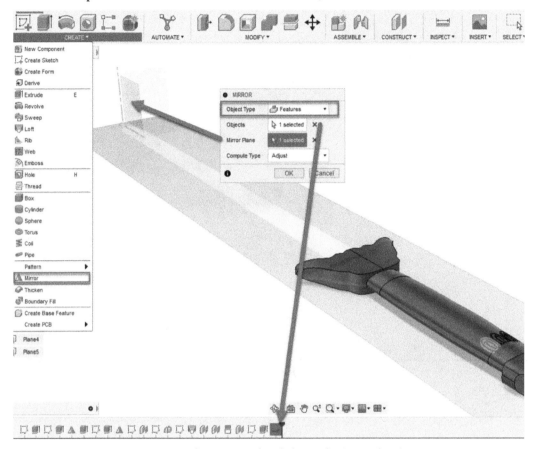

Figure 14.56 – The Mirror tool and plane selection and options

11. Now, we can mirror everything that we created for the opposite end to the guard. Turn on the midplane that we created in an earlier step, located in the `Construction` folder (my plane was labeled as **Plane2**). Now, click on the **Mirror** tool once more and select multiple features in the **Timeline** by holding down the *Ctrl* key. Select the midplane as **Mirror Plane**, and then click **OK**.

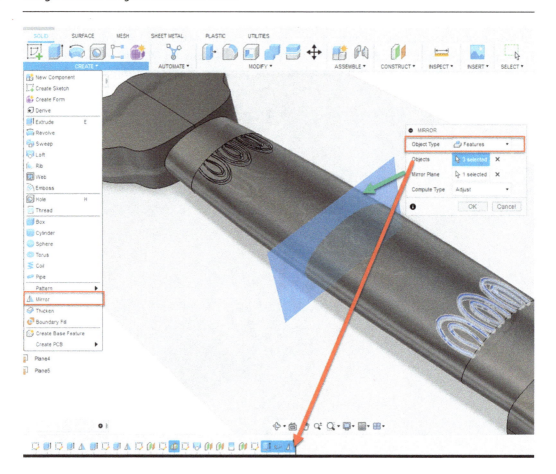

Figure 14.57 – The Mirror tool used on multiple features

This demonstrates that you don't have to create multiple sketches to create an object. Try your best to create something once and then mirror it. Fusion 360 can calculate a smaller sketch that is mirrored or arrayed once better than multiple sketches and extrusions.

12. If you want to create the funky cloud in the middle, I will let you go ahead and do that, or even better, design something that you would want on your sword using what we just learned.

13. Let's finally add the fillets to our sword handle. Try to save your fillets for last, as they take some time to calculate and can sometimes get in your way while designing. Click on the **Fillet** tool located in the **MODIFY** panel and select the faces – not the edges but the faces of the guard and the hilt. The reason that we are selecting the faces is that the fillet will be applied to all the edges without us trying to select each one individually. Set the distance to 0.05 inches and click **OK**.

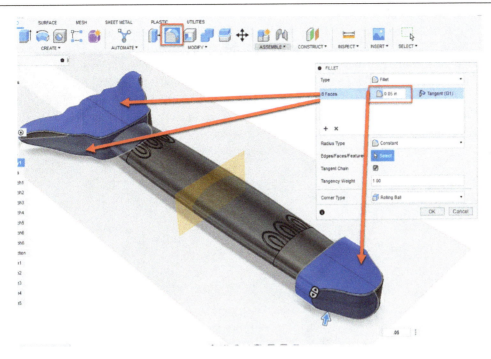

Figure 14.58 – The Fillet tool used on multiple faces

14. Turn on the sword body and turn off the Canvases folder. The sword looks great, but there is one final step. If we were to turn off the sword and orbit to see the front of the guard, we would see that there is no hole for the sword to fit into the handle. We need to create one using the **Subtract** tool.

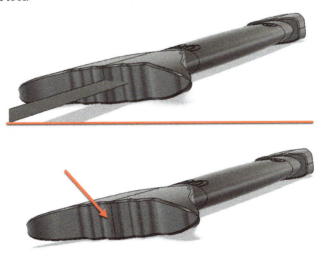

Figure 14.59 – The sword needs a handle hole

15. Return to the **Sword** eyeball icon, click on the **MODIFY** drop-down arrow, and select the **COMBINE** tool. Yes, I know what you're thinking – we are cutting the model, so why are we using the combine tool to accomplish this? It is because this is a Boolean operation that uses the **Union**, **Subtract**, and **Intersect** tools, and Fusion 360 combined them all into one command to make things easier. For **Target Body**, select the handle, as that is what we will cut. For **Tool Body**, select the sword, as that is what we will use to cut away the opening. For **Operation**, select the middle icon, which is **Cut**. Be sure to select the **Keep Tools** checkbox; otherwise, the sword will be removed from the design. Now, click **OK**.

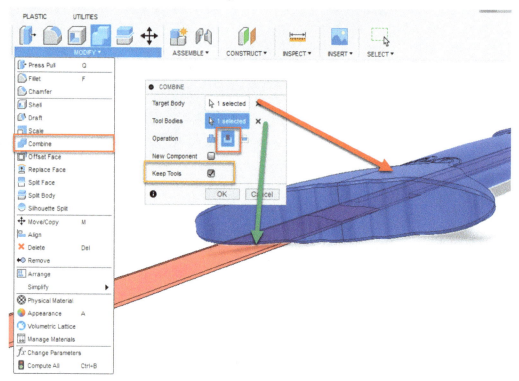

Figure 14.60 – The Combine tool with a Cut operation

16. If you turn off the sword icon now, you will see that the sword has a hole cut inside the handle.

Figure 14.61 – The handle now has a hole to fit the sword into

17. Finally, add some materials to your sword and bring it into the **RENDER** environment. Use the previous chapter tutorials as a guide if you need to.

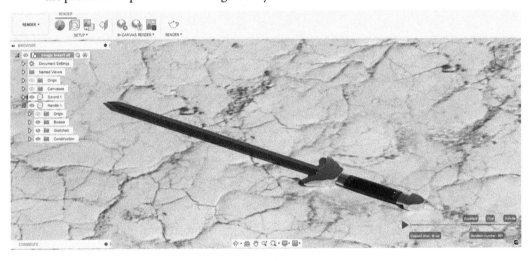

Figure 14.62 – The final sword rendering with materials

I hope your sword came out as amazing as you hoped it would! I would love to see your creations and what design you decided on for the handle. Send it to my LinkedIn account and let's share our designs.

Summary

In this chapter, we learned how to take a picture and bring it into Fusion 360 using the **Canvas** tool. We then scaled that image up to size and sketched over the image to create an amazing-looking sword. In the final chapter of the book, we will explore how to 3D-print and create a bottle topper, using the **Manufacture** tab.

15

Modeling a Bottle Topper

You've made it to the final chapter of this Fusion 360 book. I hope you have been enjoying all the projects so far. For our final chapter, we will take a scanned 3D model and import the mesh into Fusion 360. We will fix any holes within the scanned mesh and then create a bottle topper out of the mesh body. Scanning objects such as mechanical or organic parts has started to become part of the normal workday as it's much easier to pull information from the existing geometry rather than recreate it. There are a variety of free and paid products online to help get you into the scanning business for yourself. Here is a link to all3dp.com that provides a brief overview of some free and paid scanning applications: https://all3dp.com/2/best-3d-scanner-app-iphone-android-photogrammetry/.

In this chapter, we're going to cover the following main topics:

- Importing a mesh model to Fusion 360
- Fixing the mesh body
- Getting ready for 3D printing

Technical requirements

You can practice with the files provided or feel free to create your own for a more custom experience. The sample design for this chapter can be found at https://github.com/PacktPublishing/Improving-CAD-Designs-with-Autodesk-Fusion-360/tree/main/Ch15.

Importing a mesh model to Fusion 360

For the mesh model that we will be working with in this chapter, I was lucky enough to scan my father before he passed away a few years ago. We will take that original mesh and import it into Fusion 360, then scale up the model, fix any holes within the mesh, and then turn it into a bottle topper. Let's start by importing the mesh:

1. Open Fusion 360 and start a blank Fusion 360 design file. Click on the **Save** icon and create a new folder within your Packt Pub project folder and name it Ch15 Scanned Mesh. Finally, save the design file with the name Bottle Topper.

2. Download the start_Dad Model.obj mesh body from the Ch15 GitHub folder, the link to which is mentioned in the *Technical requirements* section.

3. Go to the **INSERT** panel and click on **Insert Mesh**.

Figure 15.1 – The Insert Mesh tool location

4. Locate the `start_Dad Model.obj` mesh that you downloaded from the GitHub link and click **Open**.

The mesh comes face down so we will need to rotate it to match the ViewCube with the following steps:

a. You can either use the rotational drag circles or open the **Numerical Inputs** section and type in −90 for the **X Angle** box.

b. Then, go to the **Position** area of the **INSERT MESH** floating window and click on the **Center** button icon to move the mesh to the center of the origin, and then click the **Move To Ground** icon, which will place the bottom of the mesh part on the XY plane.

c. Click **OK** to complete the **Insert Mesh** command.

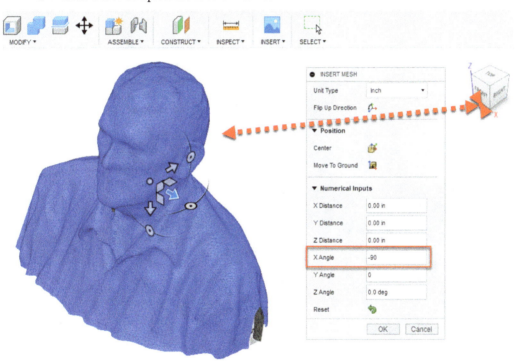

Figure 15.02 – The Insert Mesh tool options

Now that the mesh has been inserted and orientated properly, we can size the mesh body using the Scale tool.

Scaling the mesh body

Since we will be placing this mesh model on top of a bottle cap, we will need to scale the model to fit. Let's check the size of the model first by checking two points:

1. Click on the **INSPECT** panel drop-down arrow and select the **Measure** tool.

2. Click on any point on the top of the head and then click a point close to the bottom.

3. Go to the **Results** area of the **MEASURE** floating window and under **XYZ Delta**, click on the *Component 1* measurement system. You will see that the blue measurement, which represents the Z axis, says that it is currently **-0.426 in** tall. Click **Close** to complete the command.

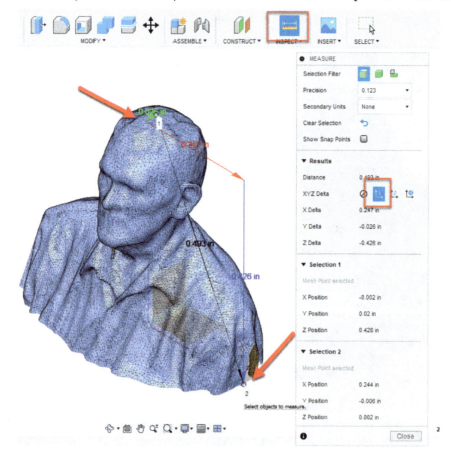

Figure 15.03 – The Measure tool location and options

A bottle cap is about 29 mm (1.14 inches) in diameter; we know this from our earlier dimensions in *Chapter 8*. So, we will need to scale the model larger, about twice the size.

4. Go to the **MODIFY** drop-down arrow and select **Scale**. The **SCALE** options floating window will open up.

5. Select the mesh body as **Entities** and for the **Point** selection, select the origin dot at the bottom where the XYZ planes meet. This will scale the model so that the bottom of the mesh stays along the XY plane.

6. Set **Scale Factor** to 2 and click **OK**.

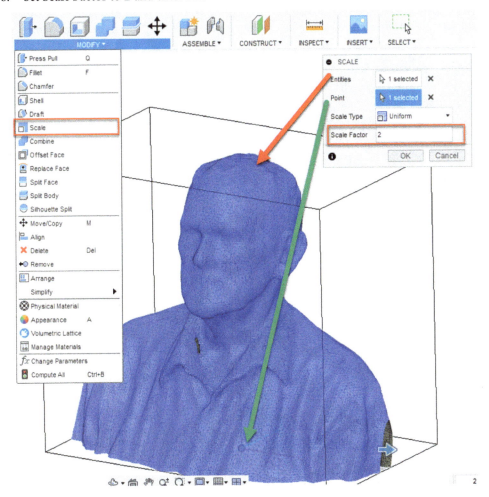

Figure 15.04 – The Scale tool location and options

Once the mesh model has been positioned on the desired location on the screen and scaled up to the approximate size, we can begin addressing any openings to establish a complete and closed mesh. We will not be able to 3D print this model without it being either a closed mesh or having added some thickness to turn it into a solid model.

Fixing the mesh body

We have a few tools at our disposal to help clean up and patch up some holes in our mesh. This is mostly what the **MESH** environment can do. It is not a buildable workspace like the other tabs, such as **SOLID** or **SURFACE**. These tools are mainly for editing purposes. Some work great while others may not work the way you had hoped. The best way to see which ones work and which don't is to test and see, so follow along with me:

1. Click on the **MESH** tab and notice that there are a lot more tools within this environment.

    Most tools are used for selecting and repairing the mesh body rather than recreating mesh surfaces. We need to first clean up the mesh by removing extra geometry that was added when my father was scanned. Orbit to view the bottom of the mesh and notice that there are some extra triangles located near the head and also near the bottom as well. These will hinder any tools such as the Thicken tool since it will confuse which way to add the thickness due to its messy geometry. We need to remove these completely using the **Direct Edit** tool.

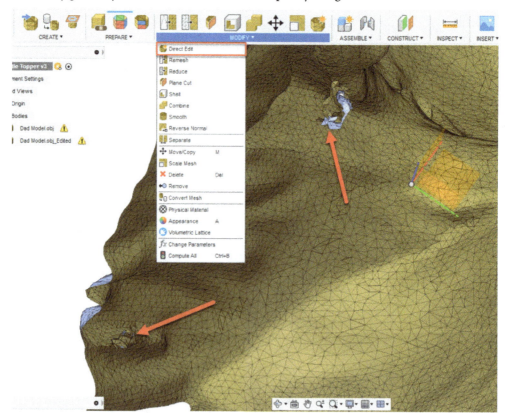

Figure 15.05 – The Direct Edit tool location and faces to remove

2. Go to **MODIFY** and click on the **Direct Edit** tool. A pop-up window will ask to select the body. Left-click anywhere on the mesh body and click **OK**.

A fly-out window will appear called **MESH SELECTION PALETTE**. This palette allows you to select mesh triangles, which you can then combine with any of the tools within the **MODIFY** drop-down arrow or hit the *Delete* key to remove some pieces. We will be selecting extra mesh triangles and hitting *Delete* to remove extra pieces. To do this, follow these steps:

a. Set **Selection Type** to **Paint**, set **Selection Filter** to **Triangles**, and set **Brush Size** to the lowest setting.

b. Now, move your mouse over the triangles that you would like to remove (use **Orbit** and **Zoom** to locate them) and left-click and drag to select as if you are painting with your mouse.

You do not have to select everything at once to remove all the triangles but, instead, take your time going around and slowly remove the extra triangle geometry. If you accidentally select triangles other than the ones you want to remove, hold down the *Alt* key and left-click to unselect them (*Figure 15.06*, green arrows).

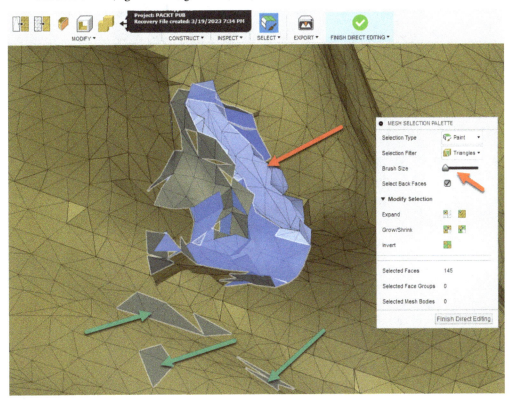

Figure 15.06 – Selecting and deselecting faces using MESH SELECTION PALETTE

3. Once you have finished cleaning up the mesh, click on the **Finish Direct Editing** button.

The hole may be too large to clean up with the **Direct Edit** and **Erase and Fill** tools so we will leave this one open for now. Look around the model for any other oddities, such as the following one with a lot of extra face edges close to a hole. There are two near the bottom that need to be cleaned up. Use **MESH SELECTION PALETTE** to clean them up as best as you can. It's okay if there are other holes; it's the extra triangle material coming out of the holes that we want to remove.

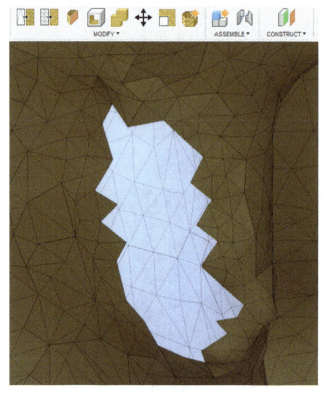

Figure 15.07 – The hole with no hanging face edges

4. Orbit to the outside of the mesh and notice that just to the right of the large hole at the top, there is a small hole that can be filled using the Direct Edit Erase and Fill tool.

5. Select the **Direct Edit** tool and click on the mesh body. Click **OK** to move on to the mesh selection palette.

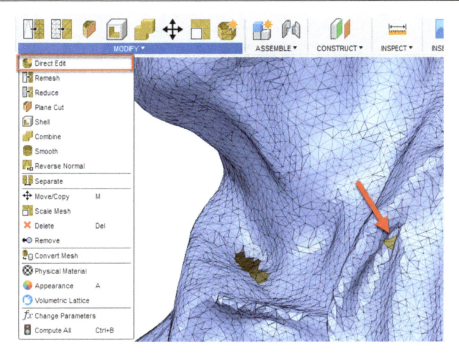

Figure 15.08 – Another hole to fill using the Direct Edit tool

6. Use the **Paint** selection tool to select the faces around the outside of the hole. Be sure not to select any other faces outside of the hole selection; if you do, hold down the *Alt* key and click to deselect.

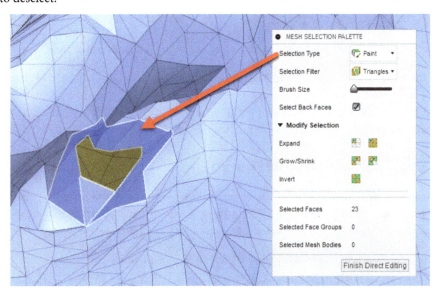

Figure 15.09 – Selecting around the edges of the hole

7. Go back up to the **MODIFY** drop-down arrow and select **Erase and Fill**.

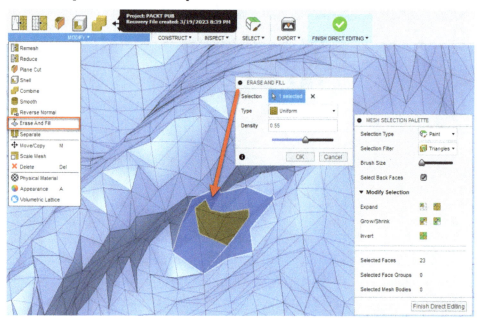

Figure 15.10 – Selecting around the edges of the hole

Notice that this tool only appears once you use the **Direct Select** tool first. Also notice that nothing happens! This may happen to you from time to time when working with meshes but not to worry, there are other tools and ways to fill in gaps. First, let's take care of the largest gap at the bottom and fill that one in the next section.

8. Click the **Finish Direct Editing** button to close out of the selection command.

Notice that there is a tool within the **PREPARE** panel called **Repair**, but this will only work on small gaps and not large holes as we have. If you try to use this tool now on the mesh, it will work but not in the way that we would like it to by closing the holes neatly. In order to fill this large hole, we will need to create a flat surface and then convert it to a mesh body, then combine the two to fill the hole.

Converting a surface to a mesh body

Since the mesh tools are typically best for fixing mechanical parts with minor errors, trying to fix an organic mesh can be a bit more complicated. Not to worry, we will create a surface shape, export it as a mesh, then reimport it to fill the bottom hole. Let's get started with the steps:

1. Click on the **SURFACE** tab and create a sketch on the XY plane.

2. Create a rectangle that extends beyond the edges of the mesh model. The size doesn't matter, just be sure that all sides of the rectangle go beyond the edges. Click on **FINISH SKETCH**.

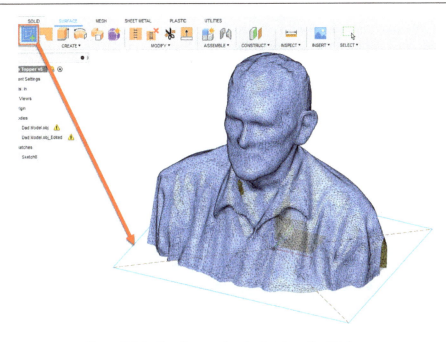

Figure 15.11 – Creating a rectangle sketch on the XY plane

3.   Within the **SURFACE** tab, click on the **Patch** tool within the **CREATE** panel. Select the rectangle sketch that was created in the previous step. Click **OK** to complete the command.

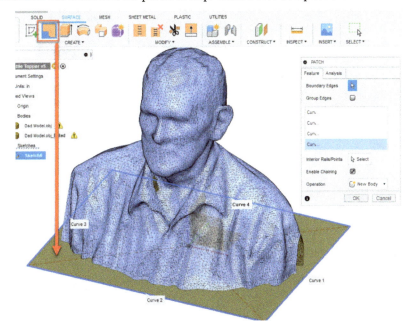

Figure 15.12 – Converting the sketch using the Patch tool

4. Within the **BROWSER** tab, right-click on the surface body that we created in the previous step and select **Save As Mesh**. Set **Format** to **STL (ASCII)** and click **OK**.

    A pop-up window will appear asking for a location to save your mesh on your computer.

5. Save it to your desktop or to the Fusion 360 cloud and give it a name such as `plane`.

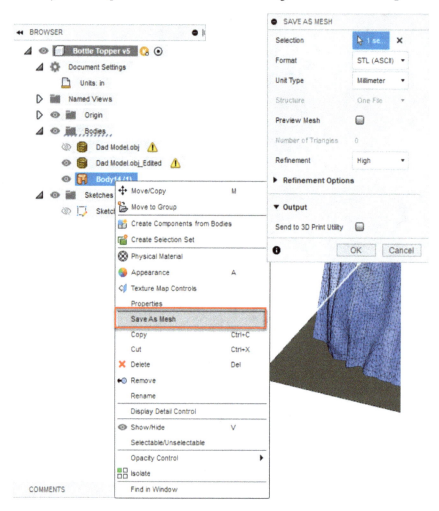

Figure 15.13 – The Save As Mesh tool

Now we will import that mesh in the same location.

6. Click on the **INSERT** drop-down arrow and select **Insert Mesh**.

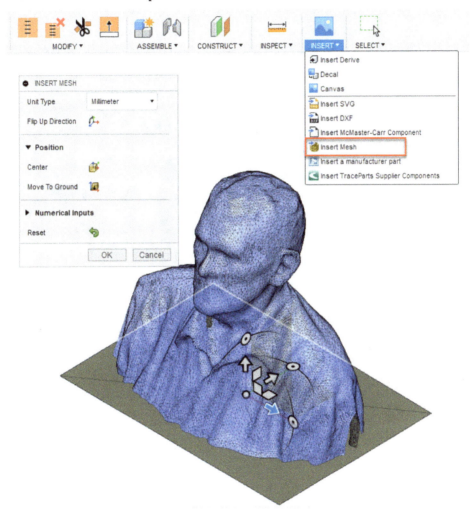

Figure 15.14 – Inserting the plane as a mesh

7. Locate where you have saved the "plane" mesh and click **OK**. The mesh inserts into the same location where it was created and at the same size (just be sure to match the mesh body units). Click **OK** to accept the defaults.

8. Within the **BROWSER**, turn off the previous **Surface** plane to see the new mesh plane.

9. Click on the **MESH** tab to go back into the Mesh environment, select the **MODIFY** drop-down arrow, and select the **Combine** tool. The **COMBINE** floating window will appear.

10. Select the head scan for **Target Body** and the plane for **Tool Bodies**. There are other Boolean operators in here such as **Union**, **Subtract**, and **Intersect** but we want the **Merge** option at the far right, which will connect the two meshes without changing the bodies. Click **OK** to complete the command.

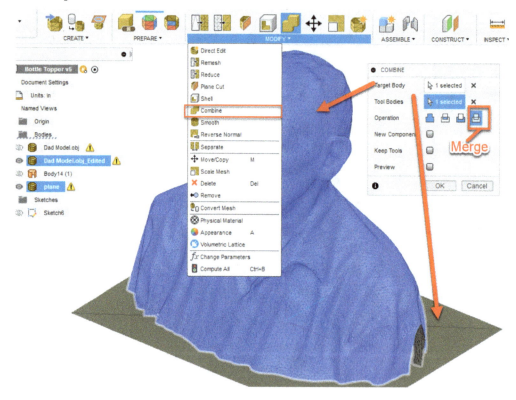

Figure 15.15 – The Combine tool with the Merge operator

Now that the large hole has been covered, we will use the **Repair** tool to fill all the remaining holes over the entire mesh body.

11. Click on the **Repair** tool located within the **PREPARE** panel. The **REPAIR** floating window will appear.

12. Select the required **Body** and set **Repair Type** to **Stitch and Remove**. This will remove any holes and then stitch up the mesh to form a complete, watertight mesh.

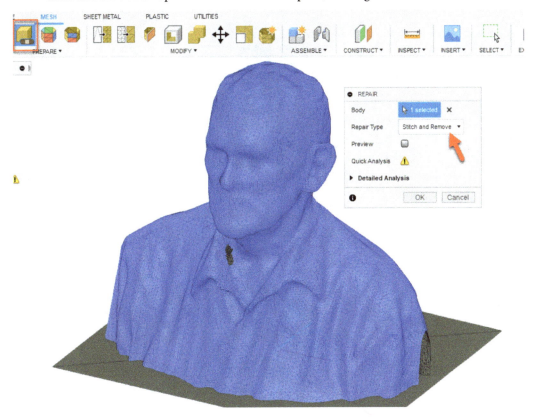

Figure 15.16 – The Repair tool with the Stitch and Remove repair type

The mesh holes have been closed but the mesh looks like it has been cut up. We can use the **Repair** tool once again to fix this.

13. Click on the **Repair** tool and select the body once again. Set **Repair Type** to **Rebuild** and then set **Rebuild Type** to **Accurate**. Click **OK** to complete the command.

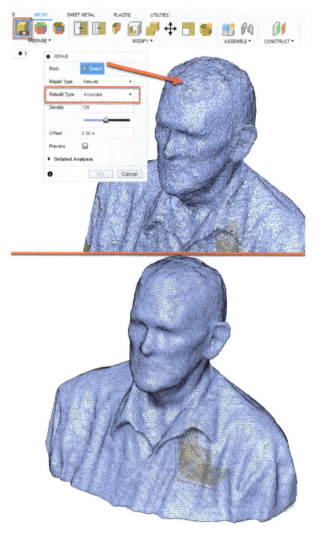

Figure 15.17 – The Repair tool with the Rebuild repair type

Amazingly, the mesh has been smoothed out and looks terrific. If you have any odd mesh geometry sticking out, use the **Mesh Edit** then **Erase and Fill** tools to remove it. Let's place this mesh on top of a bottle topper so that it can be 3D printed.

Adding a base bottle cap

We will create a simple bottle top that fits over most bottle openings and add some text such as "For Dad Only." If we look at *Chapter 8*, a small bottle cap is about 29 mm in diameter, so we will use that size for our cap model. Let's get started with the steps:

1. Click on the **SOLID** panel and create a sketch on the XY plane of a circle.

2. Set the circle to 29 mm in diameter and click on **FINISH SKETCH** to complete the command.

   Notice that I have **Units** still set to inches, but I can add dimensions in mm too. This is very handy when we are used to certain units but want to switch back and forth without any scaling problems.

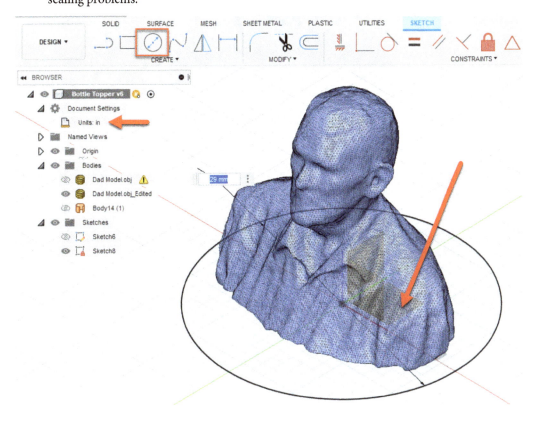

Figure 15.18 – Creating a circle sketch on the XY plane

3. Click on the **Extrude** tool and set **Distance** to **–20mm**, then click **OK**. This creates the outside of the bottle top.

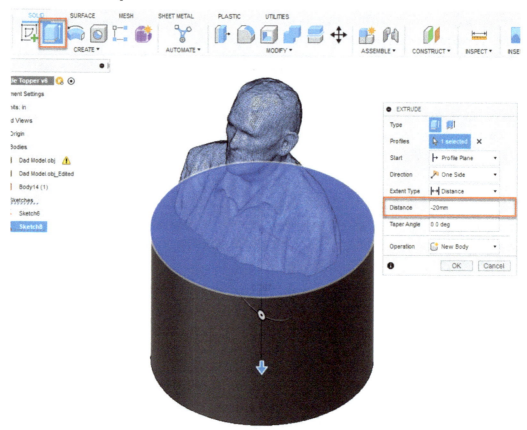

Figure 15.19 – Extruding the circle sketch

Now we need to create an opening so that it can sit on top of a bottle.

4. Select the **Shell** tool located within the **MODIFY** panel and select the bottom face of the cylinder that we just created to remove it.

5. Set **Inside Thickness** to **2.5mm** and click **OK**. This creates a thickness all around the inside that is 2.5 mm thick.

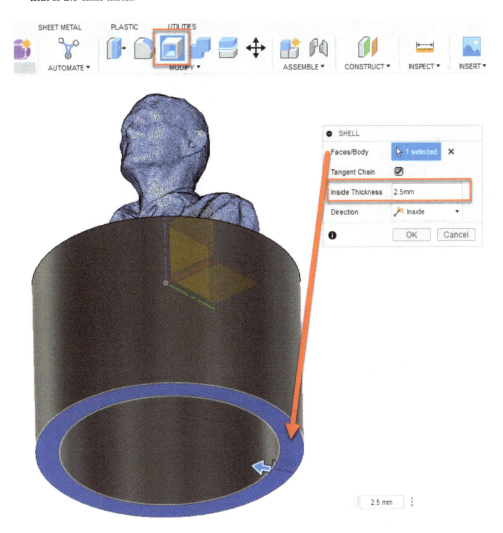

Figure 15.20 – Setting a shell for the cylinder and removing the bottom face

We will add some text to the bottle top and wrap it around the bottle cap. We will do this using the **Emboss** feature. What embossing means is raising or lowering an object, such as the text sketch, relative to another 3D model body. In order to do this, we need to first create an offset sketch.

6.   Click on the **CONSTRUCT** panel and select the **Offset Plane** tool. Select the XZ plane and set **Distance** to –**1.5** inches. Then click **OK**.

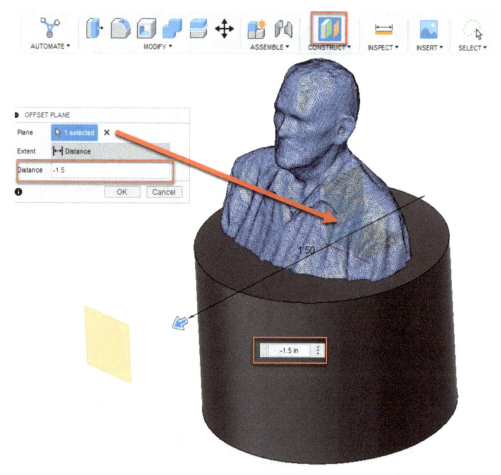

Figure 15.21 – Creating an offset plane from the XZ plane

7. Click on the **Create Sketch** tool and select the offset plane that we just created.

8. Click on the **CREATE** drop-down arrow and select the **Text** tool.

9. Drag a window from one side to the other and a floating **TEXT** window will appear. Type in some text such as FOR DAD ONLY or something similar that you would like if you are planning on 3D-printing this piece.

10. Set **Font** to **Arial Black** as we want this text to be thick when it is 3D printed, which will make it more readable.

11. Set **Height** to 0.25 inches and click **OK**.

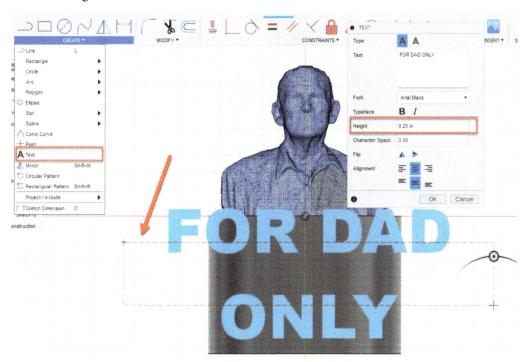

Figure 15.22 – Creating text within a sketch

We can drag the text window a little larger so that all the text fits on one line by clicking on a corner and dragging. You can also add dimensions to the exterior box frame to make sure that the text stays in its location.

12. Add dimensions as shown in *Figure 15.23* to keep the textbox from moving and then click on **FINISH SKETCH**.

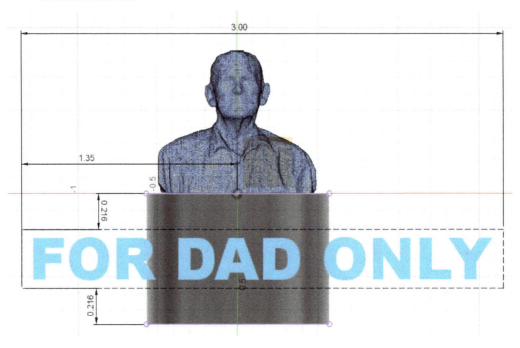

Figure 15.23 – Adding dimensions to a text frame

13. Click on the **CREATE** drop-down arrow and select the **Emboss** tool. Select the **Sketch** text as **Sketch Profile** and then select the cylinder face. The text will extend itself and wrap around the cylinder.

Amazing! You can also switch it to cut into the body by going to the **Effect** section and selecting **Cut** if you would like. Play around with some of the other settings to adjust it to your liking and then click **OK** when complete.

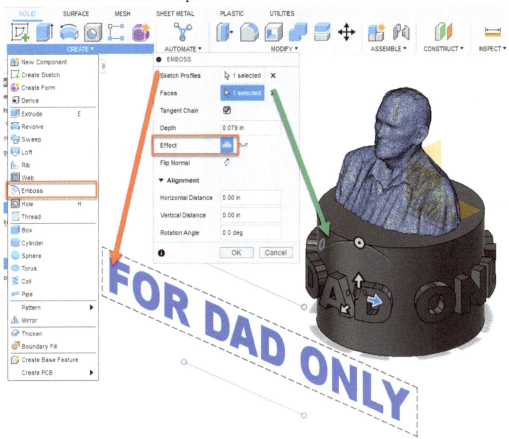

Figure 15.24 – Adding an Emboss effect to wrap text around a cylinder

The bottle topper looks great so far. Let's get it ready for 3D printing next!

Getting ready for 3D printing

The first thing we need to do is to convert the mesh into a solid body. You do have the ability to 3D print meshes without converting, but I ran into some issues when I was adding supports to this model as it treated each model separately. Converting the mesh to a solid solved this issue, so let's get into it:

1. Select the **MESH** tab and click on the **MODIFY** drop-down arrow and select the **Convert Mesh** tool. The **CONVERT MESH** floating window will open up.

2. Select the body and leave the other settings as is and click **OK**.

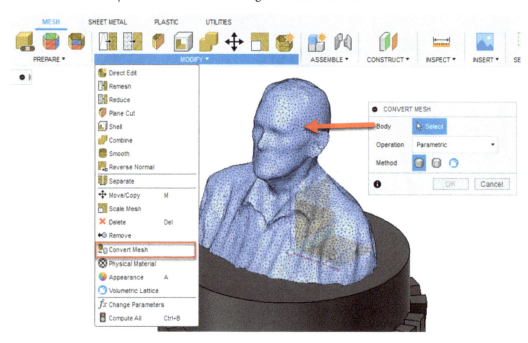

Figure 15.25 – Converting the mesh to a solid body

Now that the mesh is a solid body, we can use various solid modeling tools on it, such as the **Combine** tool to connect it with the base model. Let's first move the model down slightly so that it forms a clean edge.

3. Click on the **SOLID** tab, then select the **Move/Copy** tool located within the **MODIFY** panel.

4. Set **Move Type** to **Translate**, which will move the body from the center, and set **Z Distance** to -0.03 inches and click **OK**.

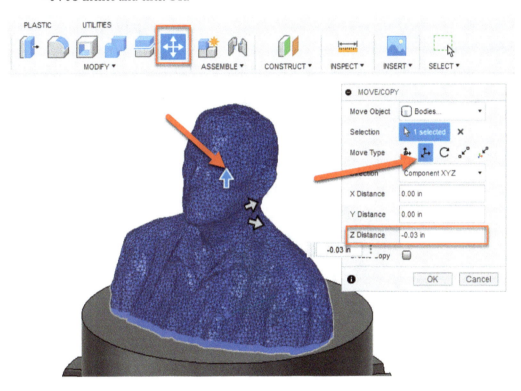

Figure 15.26 – Moving the top body in the Z direction

Now we will combine the top model and the base into one model using the **Combine** tool.

5. Click on the **MODIFY** drop-down arrow and select **Combine**. The **COMBINE** floating window will open up.

6. Set **Target Body** as the former mesh model and then select the lower base as **Tool Bodies**. Set **Operation** to **Join** and click **OK**. The models are now combined into one solid model.

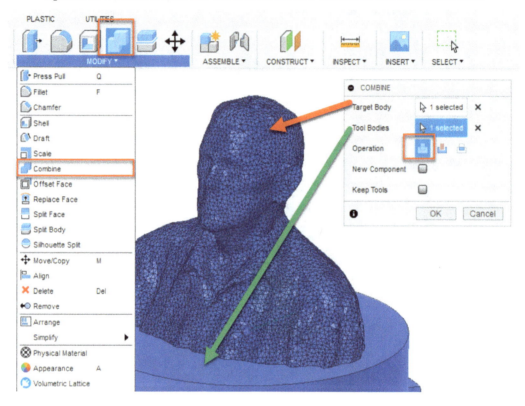

Figure 15.27 – Combining the two bodies using Join

Now that the model has been combined, let's explore the Manufacture environment and get it ready for exporting to a 3D printer.

The Manufacture environment

Fusion 360 comes with its own internal 3D printing environment within the Manufacture environment. If you wanted to print to an external printing service, you could just select the combined body that we created in **BROWSER**, then right-click on the model and select **Save As Mesh**. Doing this will open a dialog box allowing you to export the model as a mesh body, which you can send to a separate slicer program for 3D printing. Since we are working within Fusion 360 and it has its own internal slicer program, let's use this instead:

1. Click on the **DESIGN** workspace dropdown in the top-left corner of your screen and select the **MANUFACTURE** workspace.

    There are many different tabs located within this workspace for various styles of manufacturing, but the one that we want for 3D printing is **Additive**.

2. Click on the **ADDITIVE** tab and the toolbar will change to show the tools available for additive manufacturing.

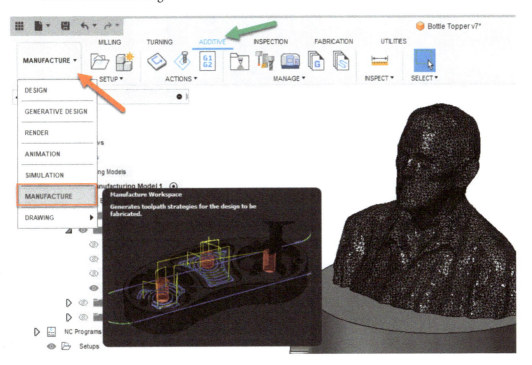

Figure 15.28 – Selecting the MANUFACTURE environment and the ADDITIVE tab

It is relatively easy to create a 3D print in Fusion as you just need to follow the toolbar from left to right starting with **SETUP**. You can run multiple different setups depending on whether you are trying various prints.

3. Click on the **SETUP** folder. The **SETUP** floating window will appear. The first thing we will need to do is select a machine. Click on the **Select** button next to **Machine**.

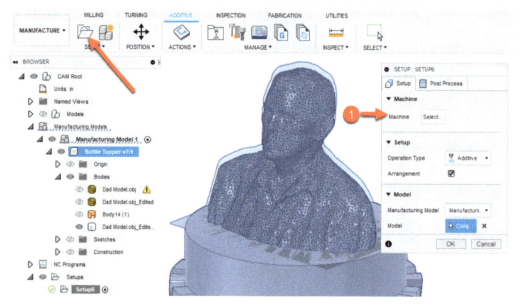

Figure 15.29 – Selecting the SETUP tool

A new floating window will appear. If you click on the Fusion 360 library drop-down arrow under the search bar to the left, you will see a large list of various 3D printer machines. If you have a 3D printer and it is not listed, you will need to locate the specs of your machine, such as the build volume, and then add it to the list as a custom machine.

4. For this tutorial, I will select **MakerBot Replicator 2X** and hit **Select**. If you have a 3D printer that you want to use and see it in that list, go ahead and select that printer as the rest of the steps are basically the same.

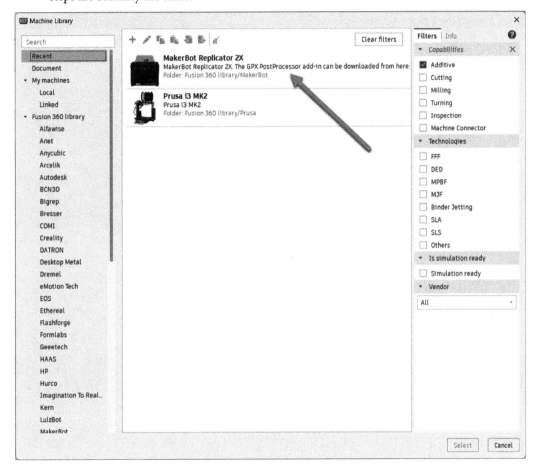

Figure 15.30 – Selecting a 3D printing machine from the machine list

Once you select the printer, you will be brought back to the **SETUP** floating window and a new selection under the **Machine** area will appear asking you to set the print settings. This is where you can select what material you will be printing with.

5.   Click on the **Print settings** button.

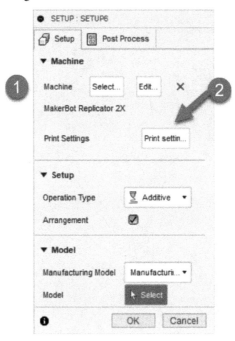

Figure 15.31 – Selecting the print settings

The **Print Setting Library** floating window will appear.

6.   Go to the Fusion 360 library search list on the left and you will see a list of materials that the printer you selected can print. I typically use **PLA 1.75mm** so let's select that one. If you are going to print with another material, feel free to select the one you wish to use.

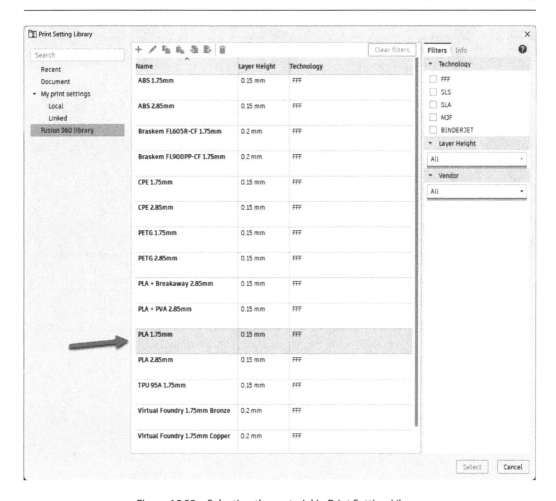

Figure 15.32 – Selecting the material in Print Setting Library

7. After selecting your material, you will be brought back out to the **SETUP** floating window. Still continuing to move down the list, leave **Setup Operation Type** as **Additive** and go down to the **Model** area.

    Your model, like mine, may have been selected already but Fusion 360 selected the entire model, which also included the original mesh, which we do not want and will cause errors if selected.

8. Click on the **X** icon to deselect what has already been selected and then click on the **Select** tool button that appears and pick the body. Now, only the body will be selected for the rest of the 3D printing setup. Click **OK** to close the **SETUP** dialog.

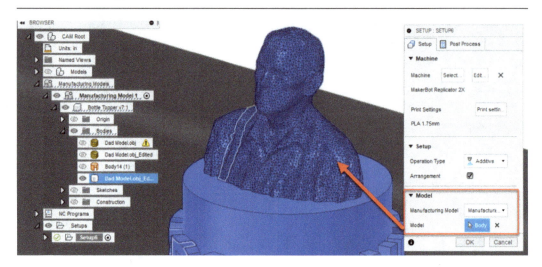

Figure 15.33 – Deselecting the entire component and selecting only the body

After you click **OK**, Fusion will process the data for a little bit and then it will show a preview of the machine you selected showing its built volume and an **ADDITIVE ARRANGE** floating window will appear. This window allows you to specify the setting of where the model will be placed and, if you had other models, how much spacing you would like to provide between them. Click **OK** since we only have one model.

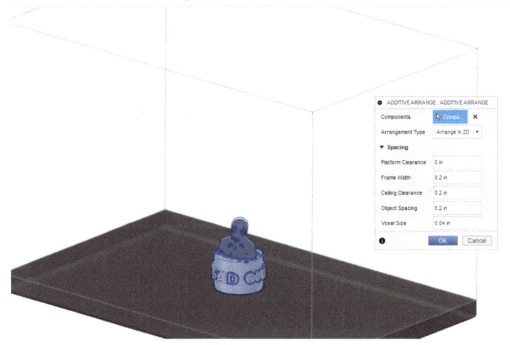

Figure 15.34 – The 3D printing machine preview and location settings

We have completed the basic setup and can now move on to the **SUPPORTS** area. If you look at the toolbar near the top of the screen now, you have some more tools to use for 3D printing. Our next step is to add support to our model. You would want to add support to any part that sticks out beyond 45 degrees as this will cause the model to drop in that location. There are two different types of support. One is **Solid Volume Support**, which is great for large spans that require extra support, and the other is **Solid Bar Support**, which is great for smaller models, like ours.

Adding supports

Supports will provide extra pillars in areas that are overhanging, such as an outstretched arm on the body of a person, for instance. This is because the printer prints in layers starting from the bottom up, and if an arm is stretching away from the body, the printer will print in an empty area and will drip printer material, causing stringing. So, to avoid any stringing and to add supports, do the following:

1. Go to the **SUPPORTS** panel and click on the **Solid Bar Support** button. The **SOLID BAR SUPPORT** floating window will appear.

    You can individually select faces by clicking faces on the body, or you can let Fusion pick what it thinks the model needs by clicking on the body within **BROWSER** on the far left.

2. In the **Supported Model** area, select the body within the **Browser Manufacturing Model 1 Bodies** folder.

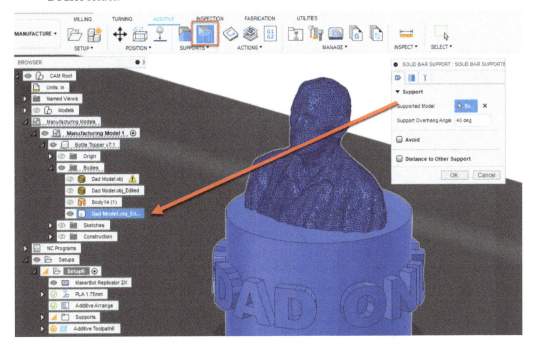

Figure 15.35 – The 3D printing machine preview and location settings

There are a lot of customized settings within this floating window, but I typically leave them as is except for one area, which is the **Bar Properties** tab.

3. Click on the **Bar Properties** tab, which is the third tab from the left, and within the **Breakpoint** area, select the drop-down arrow and choose **On part**. This thins out the top part a little so that it is easier to break off after the part has been 3D printed.

Figure 15.36 – The solid bar support bar properties breakpoint selection

A preview of the support will appear on your model. We are almost done with the setup. The final part is to simulate the toolpath.

Figure 15.37 – A preview of the printer supports

4. Click on the **Simulate Toolpath** button within the **ACTION** panel and, as long as everything has been set up correctly, your model will disappear, and some colorful lines will appear at the bottom.

    A color bar appears to the right showing the various parts of the model and a *Play* bar appears at the bottom of the screen.

5. Click on the **Play** button to see your model created one line at a time. Drag the slide bar to speed the process up a bit. Finally, if everything looks good, click on the **Post Process** button.

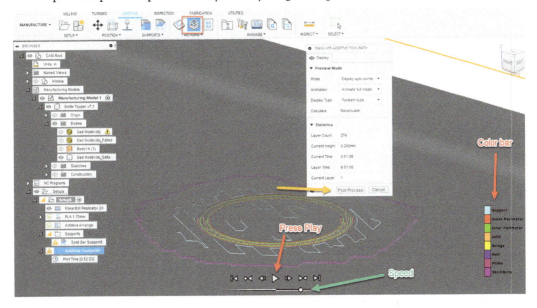

Figure 15.38 – A preview of the toolpath

Once you click on the **Post Process** button, a new dialog will open up asking where you would like to save the G-code, which you can export to your 3D printer.

6. Click on the **Name/number** area within the **Program** area and give it a name such as Bottle Topper.

7. Within the **Output folder** area, select the *folder* icon and specify a path to save the G-code to your computer.

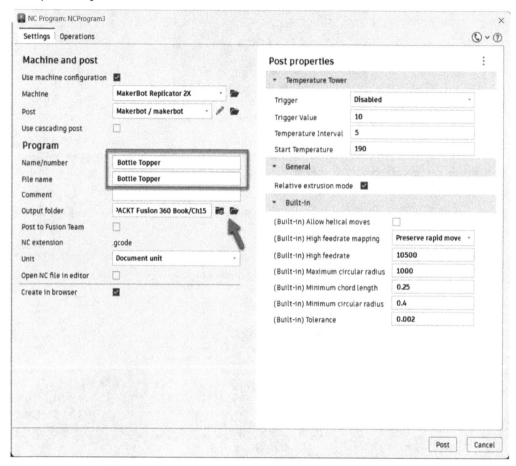

Figure 15.39 – Generating the G-code to create your 3D model

8. Click on the **Post** button and you will briefly see a *success* message appear. Now, you can copy this G-code file to your 3D printer for creation.

Congratulations on all that you have accomplished in this chapter. You can now import a 3D mesh scanned model, edit it, turn it into a solid body, and then 3D print it. If there are any problems with your 3D print such as a tight fit tolerance and the part will not connect properly, go back into Fusion and edit the model. Fusion will remember your 3D printing settings and adjust the settings so that you don't have to redo the setup again. Amazing! I hope your model turns out great!

Summary

In this chapter, we covered the basics of working with a scanned mesh model. We were able to bring the mesh into Fusion 360 and clean the model by removing any extra faces and then patching up any extra holes. We then created a solid model base and combined the two so that it could be placed into the Manufacturing space to be 3D printed.

Congratulations! You have completed this chapter and, hopefully, all the previous chapters as well. There is a lot more to accomplish using Fusion 360 and I hope this book gives you some great ideas on how to create, design, and 3D print your wonderful and amazing ideas. Please post your creations to my LinkedIn page and let me and the world see what you were able to accomplish. I'd love to see what you have done. Good luck, and I hope you have as much fun as I do while working in Fusion 360.

Index

www.packtpub.com

Subscribe to our online digital library for full access to over 7,000 books and videos, as well as industry leading tools to help you plan your personal development and advance your career. For more information, please visit our website.

Why subscribe?

- Spend less time learning and more time coding with practical eBooks and Videos from over 4,000 industry professionals

- Improve your learning with Skill Plans built especially for you

- Get a free eBook or video every month

- Fully searchable for easy access to vital information

- Copy and paste, print, and bookmark content

Did you know that Packt offers eBook versions of every book published, with PDF and ePub files available? You can upgrade to the eBook version at packtpub.com and as a print book customer, you are entitled to a discount on the eBook copy. Get in touch with us at customercare@packtpub.com for more details.

At www.packtpub.com, you can also read a collection of free technical articles, sign up for a range of free newsletters, and receive exclusive discounts and offers on Packt books and eBooks.

Other Books You May Enjoy

If you enjoyed this book, you may be interested in these other books by Packt:

Making Your CAM Journey Easier with Fusion 360

Fabrizio Cimò

ISBN: 978-1-80461-257-6

- Choose the best approach for different parts and shapes
- Avoid design flaws from a manufacturing perspective
- Discover the different machining strategies
- Understand how different tool geometries can influence machining results
- Discover how to check the tool simulation for errors
- Understand possible fixtures for raw material blocks
- Become proficient in optimizing parameters for your machine
- Explore machining theory and formulas to evaluate cutting parameters

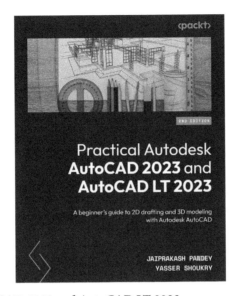

Practical Autodesk AutoCAD 2023 and AutoCAD LT 2023

Jaiprakash Pandey, Yasser Shoukry

ISBN: 978-1-80181-646-5

- Understand CAD fundamentals like functions, navigation, and components
- Create complex 3D objects using primitive shapes and editing tools
- Work with reusable objects like blocks and collaborate using xRef
- Explore advanced features like external references and dynamic blocks
- Discover surface and mesh modeling tools such as Fillet, Trim, and Extend
- Use the paper space layout to create plots for 2D and 3D models
- Convert your 2D drawings into 3D models

Packt is searching for authors like you

If you're interested in becoming an author for Packt, please visit authors.packtpub.com and apply today. We have worked with thousands of developers and tech professionals, just like you, to help them share their insight with the global tech community. You can make a general application, apply for a specific hot topic that we are recruiting an author for, or submit your own idea.

Hi!

I am Kevin Michael Land, author of *Improving CAD Designs with Autodesk Fusion 360*, I really hope you enjoyed reading this book and found it useful for kickstarting your journey of 3D design and modeling in Fusion 360.

It would really help me (and other potential readers!) if you could leave a review on Amazon sharing your thoughts on this book.

Go to the link below or scan the QR code to leave your review:

`https://packt.link/r/180056449X`

Your review will help me to understand what's worked well in this book, and what could be improved upon for future editions, so it really is appreciated.

Best Wishes,

Download a free PDF copy of this book

Thanks for purchasing this book!

Do you like to read on the go but are unable to carry your print books everywhere? Is your eBook purchase not compatible with the device of your choice?

Don't worry, now with every Packt book you get a DRM-free PDF version of that book at no cost.

Read anywhere, any place, on any device. Search, copy, and paste code from your favorite technical books directly into your application.

The perks don't stop there, you can get exclusive access to discounts, newsletters, and great free content in your inbox daily

Follow these simple steps to get the benefits:

1. Scan the QR code or visit the link below

c

2. Submit your proof of purchase
3. That's it! We'll send your free PDF and other benefits to your email directly